HAAR

38

Advances in
Polymer Science
Fortschritte der Hochpolymeren-Forschung

Editors: H.-J. Cantow, Freiburg i. Br. · G. Dall'Asta, Colleferro · K. Dušek, Prague · J. D. Ferry, Madison · H. Fujita, Osaka · M. Gordon, Colchester J. P. Kennedy, Akron · W. Kern, Mainz · S. Okamura, Kyoto C. G. Overberger, Ann Arbor · T. Saegusa, Kyoto · G. V. Schulz, Mainz W. P. Slichter, Murray Hill · J. K. Stille, Fort Collins

Polymerization Processes

With Contributions by
Al. Al. Berlin, J. M. Delvaux, N. S. Eniklopian, G. Filardo,
S. Gambino, F. Higashi, M. Kamachi, J. P. Kennedy,
G. Silvestri, S. A. Volfson and N. Yamazaki

With 33 Figures

Springer-Verlag
Berlin Heidelberg New York 1981

Editors

Prof. Hans-Joachim Cantow, Institut für Makromolekulare Chemie der Universität, Stefan-Meier-Str. 31, 7800 Freiburg i. Br., BRD

Prof. Gino Dall'Asta, SNIA VISCOSA – Centro Studi Chimico, Colleferro (Roma), Italia

Prof. Karel Dušek, Institute of Macromolecular Chemistry, Czechoslovak Academy of Sciences, 162 06 Prague 616, ČSSR

Prof. John D. Ferry, Department of Chemistry, The University of Wisconsin, Madison, Wisconsin 53706, U.S.A.

Prof. Hiroshi Fujita, Department of Polymer Science, Osaka University, Toyonaka, Osaka, Japan

Prof. Manfred Gordon, Department of Chemistry, University of Essex, Wivenhoe Park, Colchester C043SQ, England

Prof. Joseph P. Kennedy, Institute of Polymer Science, The University of Akron, Akron, Ohio 44325, U.S.A.

Prof. Werner Kern, Institut für Organische Chemie der Universität, 6500 Mainz, BRD

Prof. Seizo Okamura, No. 24, Minami-Goshomachi, Okazaki, Sakyo-Ku, Kyoto 606, Japan

Prof. Charles G. Overberger, Department of Chemistry, The University of Michigan, Ann Arbor, Michigan 48 104, U.S.A.

Prof. Takeo Saegusa, Department of Synthetic Chemistry, Faculty of Engineering, Kyoto University, Kyoto, Japan

Prof. Günter Victor Schulz, Institut für Physikalische Chemie der Universität, 6500 Mainz, BRD

Dr. William P. Slichter, Chemical Physics Research Department, Bell Telephone Laboratories, Murray Hill, New Jersey 07 971, U.S.A.

Prof. John K. Stille, Department of Chemistry, Colorado State University, Fort Collins, Colorado 805 23, U.S.A.

ISBN-3-540-10217-5 Springer-Verlag Berlin Heidelberg New York
ISBN-0-387-10217-5 Springer-Verlag New York Heidelberg Berlin

Library of Congress Catalog Card Number 61-642

This work is subject to copyright. All rights are reserved, whether the whole or part of the material is concerned, specifically those of translation, reprinting, re-use of illustrations, broadcasting, reproduction by photocopying machine or similar means, and storage in data banks. Under § 54 of the German Copyright Law where copies are made for other than private use, a fee is payable to the publisher, the amount to the fee to be determined by agreement with the publisher.

© by Springer-Verlag Berlin Heidelberg 1981
Printed in Germany

The use of general descriptive names, trademarks, etc. in this publication, even if the former are not especially identified, is not to be taken as a sign that such names, as understood by the Trade Marks and Merchandise Marks Act, may accordingly be used freely by anyone.
Typesetting and printing: Schwetzinger Verlagsdruckerei GmbH. Bookbinding: Brühlsche Universitätsdruckerei, Gießen.
2152/3140 – 543210

Table of Contents

New Condensation Polymerizations by Means of Phosphorus Compounds
N. Yamazaki and F. Higashi 1

Electrochemical Production of Initiators for Polymerization Processes
G. Silvestri, S. Gambino and G. Filardo 27

Influence of Solvent on Free Radical Polymerization of Vinyl Compounds
M. Kamachi 54

Kinetics of Polymerization Processes
Al. Al. Berlin, S. A. Volfson and N. S. Eniklopian 89

Synthesis, Characterization and Morphology of Poly(butadiene-g-Styrene)
J. P. Kennedy and J. M. Delvaux 141

Author Index Volumes 1–38 165

New Condensation Polymerizations by Means of Phosphorus Compounds

Noboru Yamazaki[1] and Fukuji Higashi[2]

[1] Department of Polymer Science, Tokyo Institute of Technology, Ohokayama, Meguro-ku, Tokyo 152, Japan
[2] Faculty of Engineering, Tokyo University of Agriculture and Technology, Koganei-shi, Tokyo 184, Japan

High-energy phosphonium compounds were prepared by the oxidation of phosphorous acids and its esters with mercuric salts or halogens in pyridines, by a hydrolysis-dehydration reaction of diphenyl and triaryl phosphites or phosphonites with pyridines and also with metal salts, or by the reaction of phosphorus halides with tertiary amines. These salts are very reactive to nucleophiles, activating carboxyl, amino, or hydroxyl compounds *via* the corresponding phosphonium salts to yield carboxylic amides and esters in high yields on further aminolysis, alcoholysis, and acidolysis. These reactions, especially the hydrolysis-dehydration reactions with phosphites, were successfully extended to the direct polycondensation reaction of dicarboxylic acids with diamines, of free α-amino acids or dipeptides, and of carbon dioxide with diamines under mild conditions, yielding linear polymers of high molecular weight (polyamides, polypeptides, polyureas). Surprisingly high molecular weight was given by the reaction in the presence of polymer matrixes. The direct polyesterification of hydroxybenzoic acids and between dicarboxylic acids and bisphenols was achieved by the phosphonium salts derived from phosphorus halides such as hexachlorocyclotriphosphatriazene.

Table of Contents

1	Introduction	2
2	**Reactions of N-Phosphonium Salts of Pyridines**	2
2.1	Oxidation-Reduction Condensations	2
2.2	Hydrolysis-Dehydration Reactions	3
2.3	Miscellaneous Reactions	6
3	**Polycondensation Reactions**	6
3.1	Polyamides	6
3.1.1	Phosphite-Pyridine Systems	6
3.1.2	Additive Effects of Functional Polymers	10
3.1.3	Phosphite-Lithium Chloride System	14
3.1.4	Matrix Polycondensations	16
3.2	Polypeptides	18
3.3	Polyureas and Polythioureas	20
3.4	Polyesters	22
4	**References**	24

1 Introduction

Adenosine triphosphate (ATP), a typical high-energy compound in living cells, plays an important role as an energy source in the production of lipids, proteins, and carbohydrates. It is consumed and regenerated *via* coupling to the phosphagen-ATP system[1]:

$$\begin{array}{c} NH_2-C-R \\ \| \\ NH \end{array} \rightleftharpoons ATP \rightleftharpoons \text{Energy}$$
$$\begin{array}{c} {}^{2-}O_3P-NH-C-R \\ \| \\ NH \end{array} \rightleftharpoons ADP$$

Phosphagen

$$\left(\begin{array}{l} \text{Creatine}; R = -N-CH_2COOH \\ \phantom{\text{Creatine}; R = } | \\ \phantom{\text{Creatine}; R = } CH_3 \\ \\ \text{Arginine}; R = -NH-(CH_2)_4-CH-COOH \\ \phantom{\text{Arginine}; R = -NH-(CH_2)_4-} | \\ \phantom{\text{Arginine}; R = -NH-(CH_2)_4-} NH_2 \end{array} \right.$$

Thus, energy obtained from reactions such as the oxidative breakdown of glucose is transferred to adenosine diphosphate (ADP), resulting in ATP, and the energy in ATP, in turn, is reserved in the phosphate bond of phosphagens.

During the course of studying chemical reactions *via* a process similar to the energy transfer in living cells described above, we have developed a new process like the phosphagen-ATP system which involves the oxidation of phosphorous acid and its esters with mercuric salts of halogens (oxidation-reduction condensation), or dephenoxylation of phosphites (hydrolysis-dehydration condensation), giving rise to the high-energy phosphonium compounds, N-phosphonium salts of pyridines.

This article reviews studies on the reactions of N-phosphonium salts of pyridines, and applications of the reactions to polymer synthesis. These studies have been carried out in the authors' laboratories since 1971.

2 Reactions of N-Phosphonium Salts of Pyridines

2.1 Oxidation-Reduction Condensations

The N-phosphonium salts of pyridines given by the oxidation of phosphorous acid and its esters are very reactive to nucleophiles, activating carboxy, amino, or hydroxy compounds to yield the corresponding carboxylic amides and esters in high yields on further aminolysis, alcoholysis, and acidolysis[2].

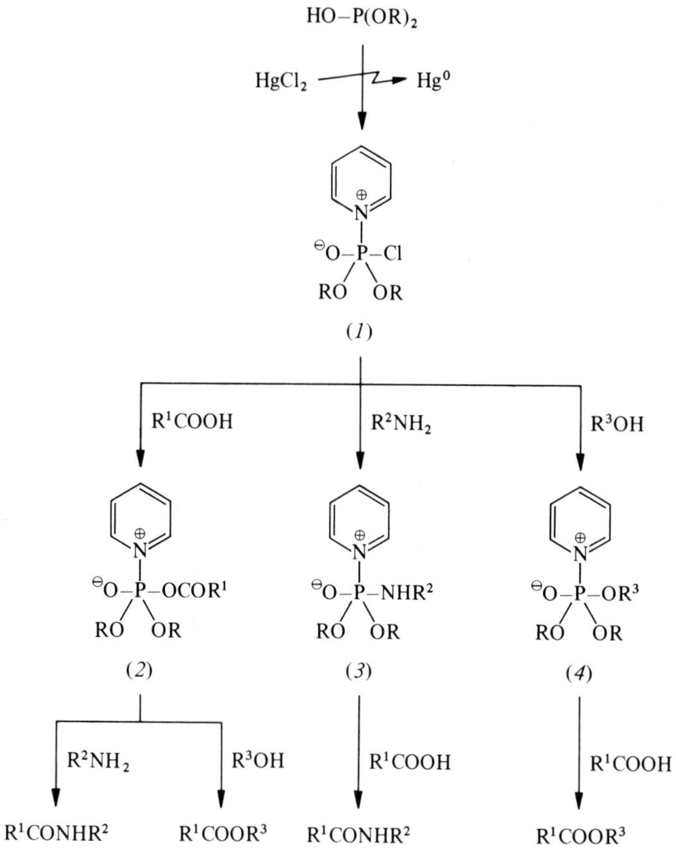

Scheme 1

```
                    HO–P(OR)₂
         HgCl₂  ———→|←——  Hg⁰
                        ↓
                    [pyridine-N⁺]
                    ⁻O–P–Cl
                    RO   OR
                       (1)
    ┌───────────────┼───────────────┐
  R¹COOH          R²NH₂           R³OH
    ↓               ↓               ↓
[pyridine-N⁺]  [pyridine-N⁺]  [pyridine-N⁺]
⁻O–P–OCOR¹    ⁻O–P–NHR²      ⁻O–P–OR³
RO   OR        RO   OR        RO   OR
  (2)            (3)            (4)
    │              │              │
  ┌─┴─┐            │              │
R²NH₂ R³OH      R¹COOH         R¹COOH
  ↓    ↓           ↓              ↓
R¹CONHR² R¹COOR³ R¹CONHR²      R¹COOR³
```

Scheme 1

These reactions were studied in terms of steric effect, acidity and basicity of carboxylic acids, amines, and tertiary amines (e.g. pyridine) by using phosphorous acid and its mono-, di-, and tri-esters, as proposed in Scheme 1 for the case of the reaction with diesters. These N-phosphonium salts (*1–4*) were separated and characterized on the bases of the IR spectra, acid-base titration, and their reactions.

For the preparation of peptides and active esters of amino acids with no detectable amounts of racemization, the reactions have successfully been extended [3].

2.2 Hydrolysis-Dehydration Reactions

We have also shown that phosphites[4], especially diphenyl and triaryl phosphites react non-oxidatively with carboxylic acids in the presence of pyridine to give acyloxy N-phosphonium salt of pyridine (*5* and *6* in Scheme 2) accompanied by dephenoxylation, which produces the corresponding amides and esters on aminolysis and alcoholysis. The structure of N-phosphonium salts such as *5* and *6* is presumed to result from the stoichiometric relationship among phosphites, pyridine, and the carboxy component.

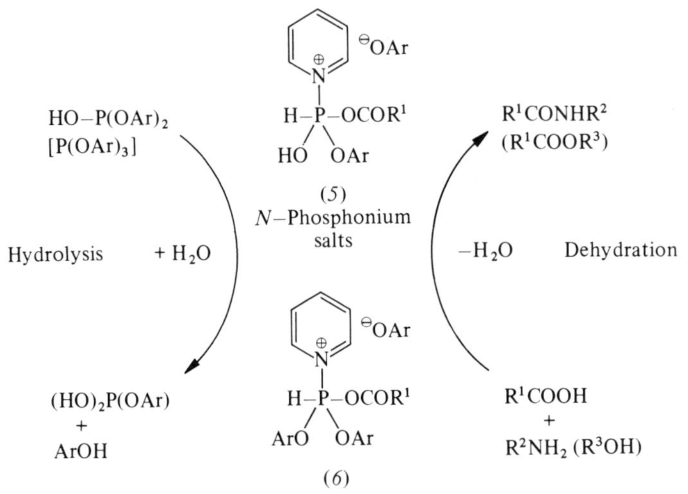

Scheme 2

In these reactions, hydrolysis of diphenyl and triaryl phosphites to monoaryl phosphites and phenol was coupled by dehydration between carboxylic acids and amines or alcohols to the corresponding amides and esters. Therefore, the reaction can be generalized as a hydrolysis-dehydration reaction (Scheme 2). The concept of the hydrolysis-dehydration reaction using phosphites has been shown to be applicable also to reactions with other phosphorus compounds such as phosphinites, phosphonites and phosphonates [5]. Aryl esters of these phosphorus compounds are effective as condensing agents in the production of carboxylic amides and esters (from carboxylic acids and amines or alcohols, respectively) whereas alkyl esters are ineffective (Eqs. (1–3)):

$$R_2P\text{-}OR + R^1COOH + R^2NH_2 \ (R^3OH) \xrightarrow{Py} R^1CONHR^2 \ (R^1COOR^3) + R_2P\text{-}OH + ROH \quad (1)$$

$$R\text{-}P(OR)_2 + R^1COOH + R^2NH_2 \ (R^3OH) \xrightarrow{Py} R^1CONHR^2 \ (R^1COOR^3) + R\text{-}P(OH)(OR) + ROH \quad (2)$$

$$R\text{-}P(O)(OR)_2 + R^1COOH + R^2NH_2 \ (R^3OH) \xrightarrow{Py} R^1CONHR^2 \ (R^1COOR^3) + ROH + R\text{-}P(O)(OR)(OH) \quad (3)$$

These reactions with phosphinites, phosphonites and phosphonates as well as phosphites have been utilized for the preparation of peptides and active esters of amino acids in good yields.

Considering that the chemical reactivity of carboxylic acids is similar to that of carbonic acids, as is observed in amide and ester formation, we have attempted the substitution of carbon dioxide for carboxylic acids in the coupling reaction with amines by using phosphites in pyridine or imidazole, and found that ureas are in fact produced in good yields (Eq. (4))[6]. Similarly, carbon disulfide reacts with amines to yield the corresponding thioureas (Eq. (5)).

$$CO_2 + 2\ RNH_2 + HO\text{-}P(OPh)_2 \xrightarrow{Py} RNHCONHR + (HO)_2\text{-}P\text{-}OPh + PhOH \quad (4)$$

$$CS_2 + 2\ RNH_2 + HO\text{-}P(OPh)_2 \xrightarrow{Py} RNHCSNHR + HS\text{-}P(OH)(OPh) + PhOH \quad (5)$$

Based on the stoichiometric involvement of phosphites and pyridine in the reaction, the latter is proposed to proceed via a carbamyl N-phosphonium salt of pyridine (7 in Scheme 3).

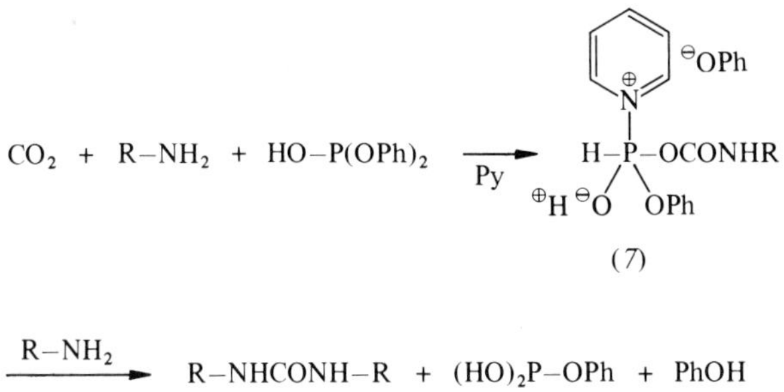

Scheme 3

In addition, we have applied successfully the concept of hydrolysis-dehydration reaction with phosphorus compounds to the reaction with sulfur compounds such as sulfites[7] as shown below:

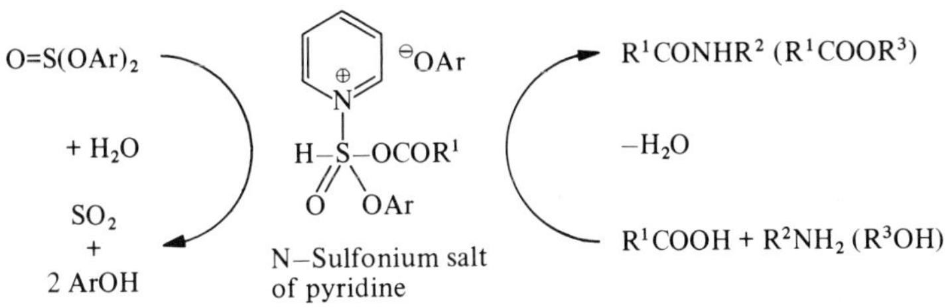

Recently, we have found other types of the hydrolysis-dehydration reactions between carboxylic acids and amines or alcohols promoted by triphenyl phosphite and alkyl halides (Eq. (6))[8] and metal salts like LiCl (Eq. (7))[9]:

$$P(OPh)_3 + R^1X \longrightarrow R^1\text{-}\overset{\oplus}{P}(OPh)_3\ X^{\ominus} \xrightarrow{R^3COOH + R^4NH_2} \overset{R^1}{\underset{}{O=P(OPh)_2}} + R^3CONHR^4 + PhOH \quad (6)$$

$$P(OPh)_3 + LiCl \longrightarrow (PhO)_3 \overset{\oplus}{\underset{Cl^{\ominus}}{P}}{-}Li \xrightarrow{RCOOH} \underset{OCOR}{\overset{\overset{\ominus}{OPh}}{H-\overset{\oplus}{P}(OPh)_2}} \quad (7)$$

$$(8) \qquad\qquad (9)$$

$$\xrightarrow{R'NH_2} RCONHR' + O=\overset{H}{P}(OPh)_2 + PhOH$$

2.3 Miscellaneous Reactions

Phosphorus halides also form reactive N-phosphonium salts which activate amino groups to yield ureas and carboxamides on treatment with carbon dioxide[10] and carboxylic acids[11]. Among phosphorus halides hexachlorocyclotriphosphatriazene[a] has proved to be useful reagent for the preparation of carboxylic amides[12] and esters[13], and has been extended to the direct polyesterification of hydroxy benzoic acids as described in Sect. 3.4.

3 Polycondensation Reactions

Since the reactions with phosphorus compounds give relatively high yields of carboxylic amides and esters and ureas, they have been extended to the direct polycondensations of dicarboxylic acids with diamines and bisphenols, of free α-amino acids or dipeptides, and of carbon dioxide and carbon disulfide with diamines under mild conditions to give polyamides, polyesters, polyureas, and polythioureas.

3.1 Polyamides

3.1.1 Phosphite-Pyridine Systems

Direct polycondensations of aromatic diamines with dicarboxylic acids have generally been described as a poor route to high molecular linear polyamides. Recently, high molecular weight polyamides have been obtained with limited success by a melt polymerization of 4,4'-diaminodiphenylmethane (MDA) with aliphatic dicarboxylic acids[14].

Surprisingly, by using the reactions via 5 and 6 in Scheme 2, polyamides of high molecular weight are obtained directly from dicarboxylic acids and diamines in N-methylpyrrolidone (NMP) solutions containing pyridine (Table 1)[15]. A combination of aromatic diamines with aliphatic dicarboxylic acids gives polymers of higher viscosity than the use of an aliphatic diamine alone. On the other hand, 4,4'-diaminodiphenylsulfone yields low viscous polymers, probably because of the lower basicity.

---

[a] 2, 2, 4, 4, 6, 6-Hexachloro-1, 3, 5, 2 λ^5, 4 λ^5, 6 λ^5-triazatriphophorine

Table 1. Direkt synthesis of polyamides by using phosphites and phosphonites in NMR-Py solution[a]

Dicarboxylic acid	Diamine	HO–P(OPh)$_2$ Yield (%)	η_{inh}	P(OPh)$_3$ Yield (%)	η_{inh}	Et–P(OPh)$_2$ Yield (%)	η_{inh}
HOOC–(CH$_2$)$_4$–COOH	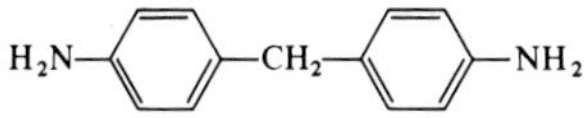 H$_2$N–C$_6$H$_4$–CH$_2$–C$_6$H$_4$–NH$_2$	94	1.22	100	0.63	97	0.97
	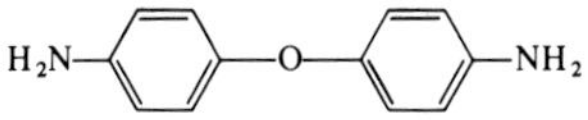 H$_2$N–C$_6$H$_4$–O–C$_6$H$_4$–NH$_2$	100	0.95	100	0.74	97	1.65
	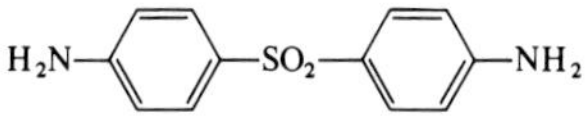 H$_2$N–C$_6$H$_4$–SO$_2$–C$_6$H$_4$–NH$_2$	86	0.28	78	0.18	62	0.22
	 H$_2$NCH$_2$–C$_6$H$_4$–CH$_2$NH$_2$	73	0.22	85	0.31	85	0.26
HOOC–(CH$_2$)$_8$–COOH	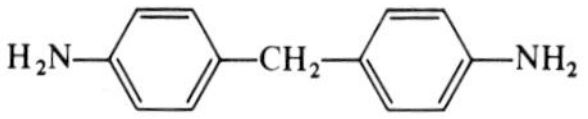 H$_2$N–C$_6$H$_4$–CH$_2$–C$_6$H$_4$–NH$_2$	97	1.45	100	1.07	96	1.12
	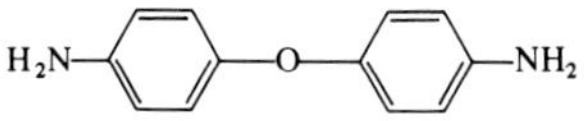 H$_2$N–C$_6$H$_4$–O–C$_6$H$_4$–NH$_2$	–	–	100	1.84	98	1.57
 HOOC–C$_6$H$_4$–COOH (para)	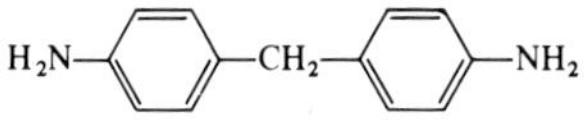 H$_2$N–C$_6$H$_4$–CH$_2$–C$_6$H$_4$–NH$_2$	100	0.23	100	0.34	98	0.30
	 H$_2$N–C$_6$H$_4$–NH$_2$	100	0.12	–	–	–	–
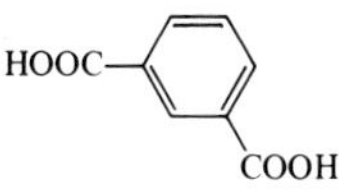 HOOC–C$_6$H$_4$–COOH (meta)	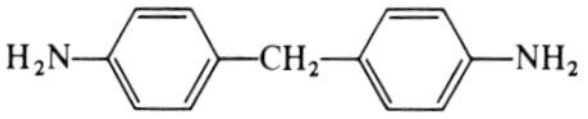 H$_2$N–C$_6$H$_4$–CH$_2$–C$_6$H$_4$–NH$_2$	–	–	100	0.54	98	0.43
	 H$_2$N–C$_6$H$_4$–NH$_2$	84	0.26	–	–	–	–
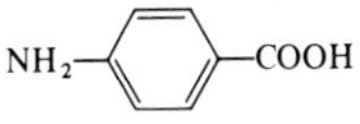 NH$_2$–C$_6$H$_4$–COOH		98	0.16	100	0.22	97	0.21

[a] [Monomer] = 0.25 mol/l; [HO–P(OC$_6$H$_5$)$_2$] = [P(OC$_6$H$_5$)$_3$] = [C$_2$H$_5$–P(OC$_6$H$_5$)$_2$] = 1.0 mol/mol of monomer; solvent = NMP/Py = 40/10 (ml/ml); temperature = 100 °C; reaction time = 6 h

Aromatic dicarboxylic acids, even with aromatic diamines and aromatic amino acids, do not form high viscous polymers. Isophthalic acid (IPA) gives higher viscous polymers than terephthalic acid. From this result it may be concluded that polycondensation is favored with higher solubility of polymer.

The difficulty of obtaining aromatic polyamides as shown above is overcome to a large extent by carrying out the polycondensation reaction in the presence of metal salts capable of improving the dissolving power of the polyamides[16]. As Table 2 shows, the addition of LiCl or CaCl$_2$ to the reaction mixture favors the polycondensation of p-aminobenzoic acid(p-ABA), giving high molecular weight poly-p-benz-

Table 2. Polycondensation reaction of p-ABA in the presence of several metal salts[a]

Metal salt	Polymer Yield (%)	η_{inh}[b]
LiCl	100	1.27
LiOCOCH$_3$	0	–
CaCl$_2$	98	1.07
CaCl$_2 \cdot$ 2H$_2$O	71	0.04
KSCN	46	0.12
MgCl$_2$	100	0.31
ZnCl$_2$	97	0.20
None	100	0.22

[a] [Monomer] = 0.4 mol/l; [P(OC$_6$H$_5$)$_3$] = 1.0 mol/mol of monomer; [metal salt] = 4 wt-% in the solvent; solvent = NMP/Py = 40/10 (ml/ml); temperature = 100 °C; time = 6 h
[b] Measured in H$_2$SO$_4$ at 30 °C

amide in quantitative yield. There are observed maxima of molecular weight of polymer at a concentration of about 4 wt-% of LiCl or 8 wt-% of CaCl$_2$ in the reaction mixture. Further addition of metal salt retards polycondensation and almost no polymer is obtained in the presence of more than 12 wt-% LiCl or 20 wt-% CaCl$_2$, where the reaction mixtures are deeply colored.

Considering that the presence of 2 wt-% LiCl and 5 wt-% CaCl$_2$, corresponding to an equivalent of phosphite and p-ABA, is very effective, the salts might participate in the reaction itself. Therefore, metal salts may contribute to the improvement of the dissolving power of the resulting polyamide and also to the suppression of the side reaction owing to the formation of complexes between phenol (derived from phosphite) and metal salts, such as those of CaCl$_2$ with alcohols and phenols.

It was expected that the polycondensation reaction at high temperatures might favor the solubility of the resulting polymer, but be undesirable for the stability of the complexes of phenol with metal salts. As a consequence, an optimum of the reaction temperature might be required in the polycondensation reaction.

An optimum of viscosity (η_{inh} = 1.71) is observed at a reaction temperature at around 80 °C in the polycondensation of p-ABA. Above this temperature the viscosity decreases gradually with temperature. Only low-viscosity polymers are obtained at 60 °C.

Of the solvents tested in the polycondensation of p-ABA in NMP/pyridine, NMP has been found to be most effective and N,N-dimethylacetamide (DMAc), in which the reaction mixture becomes light yellow gives rise to moderate yields whereas dimethylformamide (DMF) largely retards the reaction, probably because of a side reaction of DMF with LiCl at high temperatures. This is indicated by deep discoloration of the reaction mixture.

The viscosity of the polymer varies with the amount of pyridine in the NMP-pyridine mixed solvent, showing the highest value at relatively high pyridine content

(40%) in spite of unfavorable results in pyridine alone. This result suggests that the solvent of this composition has a strong solvating power, as a combination of NMP and hexamethyl-phosphoric triamide (HMPA) containing LiCl, each of which alone fails to dissolve poly-p-benzamide, is a very powerful solvent.

Several wholly aromatic polyamides have been prepared by use of triphenyl phosphite in NMP-pyridine solution containing 4 wt-% LiCl (Table 3). The combination of isophthalic acid with diamines gives a polymer of high viscosity whereas terephthalic acid (TPA) with pK_a values similar to those of isophthalic acid does not yield high-viscosity polymers.

Table 3. Preparation of aromatic polyamides by means of triphenyl phosphite in NMP-pyridine solution containing 4 wt-% LiCl[a]

Dicarboxylic acid[b]	Diamine	Polymer η_{inh}[c]
HOOC—⌬—COOH (3.72, 4.40)	H₂N—⌬—NH₂	1.14
	H₂N—⌬—O—⌬—NH₂	1.34
	H₂N—⌬—CH₂—⌬—NH₂	0.93
HOOC—⌬—COOH (3.54, 4.46)	H₂N—⌬—NH₂	0.19 (0.21)[d]
	H₂N—⌬—O—⌬—NH₂	0.32
	H₂N—⌬—CH₂—⌬—NH₂	0.33
H₂N—⌬—COOH (3.07, 4.70)		1.32
H₂N—⌬—COOH (2.28, 4.89)		0.43

[a] [Monomer] = 0.6 mol/l; [P(OC₆H₅)₃] = 1.0 mol/mol of monomer; solvent = NMP/Py = 20/15 (ml/ml); temperature = 100 °C; time = 3 h
[b] Values in parentheses are pK_1 and pK_2
[c] Polymers were obtained in quantitative yields, and the viscosity was measured in H_2SO_4 at 30 °C
[d] CaCl₂ (8 wt-%) was used in place of LiCl

3.1.2 Additive Effects of Functional Polymers

In the studies on polycondensation reactions, we have found that high molecular weight poly(terephthalamides) are obtained by use of poly(4-vinylpyridine) (P4VP) in place of pyridine, and the amount and the molecular weight of P4VP added and of solvent systems affect the molecular weight of the resulting polyamides[17]. The inherent viscosity of the resulting polymers increases with increasing the added amount of P4VP, a maximum value being given by nearly equimolar amount of the polymer to TPA.

$$HOOC-R-COOH \xrightarrow[NMP/LiCl]{P(OC_6H_5)_3/P4VP}$$

(structure 10)

$$\xrightarrow{NH_2-R'-NH_2} +OC-R-CONH-R'-NH+_n + 2\,C_6H_5OH + 2\,HO-P(OC_6H_5)_2$$

The viscosity varied also with the amounts of pyridines, and the polymer of higher viscosity was obtained in the wide range of pyridine content (Py/P4VP = 6–26). Further amounts did not give favorable results (Table 4).

No significant effects of molecular weights of P4VP upon those of polymer produced were observed, and suggested that the polycondensation reaction promoted by P4VP may not be a matrix polycondensation in which the molecular weight of the polymer formed can be dependent on the molecular weight of polymer used as matrix.

Table 4. Polycondensation of TPA and MDA in the presence of P4VP in NMP-pyridine solvent[a]

Py (ml)	NMP (ml)	P4VP (g)	Py/P4VP (mol/unit mol)	η_{inh}[b]
5	45	1.0	6.5	1.13
10	40	1.0	13	1.04
10	40	0.5	26	1.17
20	30	1.0	26	0.67
20	30	0	–	0.33

[a] TPA = MDA = 0.01 mol; $P(OC_6H_5)_3$ = 0.022 mol; P4VP (MW) = 1.0×10^5; LiCl = 2.0 g; time = 2 h; temperature = 100 °C
[b] Measured in H_2SO_4 at 30 °C

Table 5. Preparation of various polyamides in NMP-pyridine solution in the presence of P4VP[a]

	Dicarboxylic acid	Diamine	η_{inh}[b,c]
1.	HOOC–C₆H₄–COOH	H₂N–C₆H₄–NH₂	0.45 (0.19)
2.	HOOC–C₆H₄–COOH	H₂N–C₆H₄–CH₂–C₆H₄–NH₂	1.17 (0.33)
3.	HOOC–C₆H₄–COOH	H₂N–C₆H₄–O–C₆H₄–NH₂	1.68 (0.32)
4.	HOOC–C₆H₄–COOH (m)	H₂N–C₆H₄–NH₂	0.98 (1.14)
5.	HOOC–C₆H₄–COOH (m)	H₂N–C₆H₄–CH₂–C₆H₄–NH₂	0.90 (0.93)
6.		HOOC–C₆H₄–NH₂ (p)	1.25 (1.32)
7.		HOOC–C₆H₄–NH₂ (m)	0.47 (0.44)

[a] [Monomer] = 0.4 mol/l; P(OC₆H₅)₃ = 1.1 mol/mol of monomer; P4VP (MW = 1.0 × 10⁵) = = 0.5 g; LiCl = 2.0 g; NMP/Py = 40/10 (ml/ml); temperature = 100 °C; time = 2 h
[b] Measured in H_2SO_4 at 30 °C
[c] Values in parentheses are reported in our previous paper[16] for the reaction in the absence of P4VP

Several aromatic polyamides from aromatic dicarboxylic acids and diamines were prepared by the reaction in the presence of P4VP (MW = 1.0×10^5), and the results were summarized in Table 5. Combination of TPA with diamines, especially with MDA and 4,4′-diaminodiphenylether, gave polymers of satisfactory viscosities up to five times higher than those obtained in the absence of the matrix. On the other hand, the reactions of IPA and diamines, and of *p*- and *m*-ABA were almost not affected by P4VP.

In the reaction with P4VP the matrix was assumed to be involved both in the increasing local concentration of TPA through the adsorption and in the activation *via* the N-phosphonium salt of P4VP (*10*). The addition of poly-(ethylene oxide) (PEO) instead of P4VP also increased inherent viscosity of the polymer produced[18]. PEO was expected to participate in the concentration alone by the interaction with carboxyl groups via hydrogen bonding but not in the activation.

Table 6. Effect of the amount of PEO on the polycondensation of TPA and MDA in the presence of triphenyl phosphite[a]

Amount of PEO g	wt-%	unit mol/mol of TPA	Polymer η_{inh}[b]
0	0	0	0.33
0.125	0.25	0.28	0.91
0.25	0.5	0.56	1.0
0.5	1.0	1.14	0.88
1.0	2.0	2.27	0.74
2.0	4.0	4.54	0.61

[a] TPA = MDA = 0.01 mol; P(OC$_6$H$_5$)$_3$ = 0.022 mol; PEO (MW) = 3~5 × 10^5; LiCl = 2.0 g, Py/NMP = 20/30 (ml/ml), temperature = 100 °C; time = 2 h
[b] Measured at 30 °C in H$_2$SO$_4$

Optimum η_{inh} was obtained by 0.5 wt-% of PEO, corresponding to half a unit mole equivalent of TPA (Table 6). The viscosity of polymer produced increased, though not so markedly, with molecular weight of PEO, showing maximum values by the use of PEO in the range of 2.0 × 10^4 to 5.0 × 10^5. The polycondensation reaction in the presence of PEO was affected by the amount of LiCl added, and the most favorable results were obtained by the presence of more than 6 wt-% LiCl, which was higher than about 2 wt-% for the reaction in the absence of PEO.

These better results obtained by PEO and LiCl may be due to an effective adsorption of TPA on PEO through solvated lithium ion.

$$\overset{\frown}{\underset{\smile}{O}} \rightarrow Li^{\oplus} \cdots O^{\ominus}-\underset{\underset{O}{\|}}{C}-R$$

The reaction were not enhanced by PEO in pyridine in which the concentration of TPA on PEO would be negligibly small because of exclusive trapping of TPA by pyridine. When the level of pyridine became lower and the amount of TPA on PEO

Table 7. Polycondensation reaction of TPA and MDA in various solvent systems[a]

Solvent System Py(ml)	NMP(ml)	η_{inh}
50	0	0.16
20	30	1.0
15	35	1.32
10	40	1.28
5	45	1.10

[a] TPA = MDA = 0.01 mol; P(OC$_6$H$_5$)$_3$ = 0.022 mol; LiCl = 2.0 g, PEO (MW = 3~5 × 10^5) = 0.25 g; temperature = 100 °C; time = 2 h

probably increased, the reaction was remarkably promoted to yield polymer of higher viscosity. However, excess adsorption of TPA on PEO appeared to be less effective for the reaction, and the best results were obtained in the solvent at relatively higher level (30–40%) of pyridine whose composition may give the most favorable concentration of TPA on PEO (Table 7).

Table 8. Preparation of various polyamides in NMP-pyridine solution in the presence of PEO[a]

	Dicarboxylic Acid	Diamine	η_{inh}[b,c]
1.	HOOC–C₆H₄–COOH (para)	H₂N–C₆H₄–NH₂	0.63 (0.19)
2.		H₂N–C₆H₄–CH₂–C₆H₄–NH₂	1.10 (0.33)
3.		H₂N–C₆H₄–O–C₆H₄–NH₂	1.88 (0.32)
4.		H₂N–C₆H₄–S–C₆H₄–NH₂	0.89
5.		H₂N–C₆H₄–SO₂–C₆H₄–NH₂	0.39
6.	HOOC–C₆H₄–COOH (meta)	H₂N–C₆H₄–NH₂	1.06 (1.14)
7.		H₂N–C₆H₄–CH₂–C₆H₄–NH₂	1.03 (0.93)
8.		H₂N–C₆H₄–O–C₆H₄–NH₂	1.07 (1.34)
9.	HOOC–C₆H₄–NH₂ (para)		1.20 (1.32)
10.	HOOC–C₆H₄–NH₂ (meta)		0.38 (0.44)

[a] [Monomer] = 0.4 mol/l; [P(OC₆H₅)₃] = 1.1 mol/mol of monomer; PEO (MW = 3 ~ 5 × 10⁵) = = 0.25 g; LiCl = 3.0 g; Py/NMP = 10/40 (ml/ml); temperature = 100 °C; time = 2 h
[b] Measured in H_2SO_4 at 30 °C
[c] Values in parentheses are reported in our previous paper[16] for the polycondensation in the absence of PEO

In the preparation of several aromatic polyamides in NMP-pyridine solution containing PEO and LiCl (6 wt-%) (Table 8), similarly to the reactions in the presence of P4VP, the reactions of TPA with diamines were largely facilitated and produced polymer of high viscosity, whereas those of isophthalic acid with diamines and of m- and p-ABA did not affected by the presence of PEO, giving polymer of viscosities as high as those obtained in the absence of PEO.

3.1.3 Phosphite-Lithium Chloride System

The polycondensation reaction with triphenyl phosphite and metal salts was affected by metal salts used. Among metal salts examined, the addition of LiCl and $CaCl_2$ results in an increase of the yield and the molecular weight of the polyamide. $MgCl_2$ which could not be expected to improve the solvating power is also effective. The addition of $ZnCl_2$ and KBr can also facilitate the condensation reaction, though the results are not satisfactory. The use of $NiCl_2$ does not give rise to the formation of a polyamide. Metal halides affect the reaction, lithium chloride giving more favorable results than its bromide and fluoride, while the nitrate is ineffective in the polycondensation.

The polycondensation reaction of p-ABA has been carried out in various solvents capable of solvating LiCl, and results are compiled in Table 9, which reveals that carboxamide solvents such as NMP or N,N-dimethylacetamide are effective whereas no polyamides are obtained in DMF, DMSO or in ethers such as diglyme. It is assumed that solvents may affect the formation and the reactivity of 8 and/or 9 as described as follows (*11* and *12*):

(*11*)   (*12*)

Table 9. Solvent effect on the polycondensation of p-ABA[a]

Solvent	Yield (%)	η_{inh}
NMP	100	0.88
DMAc	100	0.40
HMPA[b]	58	0.08
DMF	0	–
DMSO[c]	0	–
Diglyme	0	–

[a] [p-ABA] = [P(OC$_6$H$_5$)$_3$] = 0.01 mol; LiCl = 1.0 g; solvent = 25 ml; temperature = 100 °C; time = 2 h
[b] HMPA = hexamethylphosphoric triamide
[c] DMSO = dimethyl sulfoxide

Table 10. Preparation of various polyamides using triphenyl phosphite in NMP containing 4 wt-% LiCl[a]

	Dicarboxylic Acid	Diamine	η_{inh}[b]
1.	HOOC–(CH$_2$)$_4$–COOH	H$_2$N–C$_6$H$_4$–NH$_2$	1.17
2.		H$_2$N–C$_6$H$_4$–CH$_2$–C$_6$H$_4$–NH$_2$	0.82
3.		H$_2$N–C$_6$H$_4$–O–C$_6$H$_4$–NH$_2$	1.45
4.	HOOC–C$_6$H$_4$–COOH (para)	H$_2$N–C$_6$H$_4$–NH$_2$	0.53
5.		H$_2$N–C$_6$H$_4$–CH$_2$–C$_6$H$_4$–NH$_2$	0.86
6.		H$_2$N–C$_6$H$_4$–O–C$_6$H$_4$–NH$_2$	1.33
7.		H$_2$N–C$_6$H$_4$–SO$_2$–C$_6$H$_4$–NH$_2$	0.49
8.	HOOC–C$_6$H$_4$–COOH (meta)	H$_2$N–C$_6$H$_4$–NH$_2$	0.74
9.		H$_2$N–C$_6$H$_4$–CH$_2$–C$_6$H$_4$–NH$_2$	0.43
10.		HOOC–C$_6$H$_4$–NH$_2$ (para)	0.88
11.		HOOC–C$_6$H$_4$–NH$_2$ (meta)	0.19

[a] [monomer] = 0.4 mol/l; [P(OC$_6$H$_5$)$_3$] = 1.0 mol/mol of monomer; temperature = 100 °C; time = 2 h
[b] Measured in H$_2$SO$_4$ at 30 °C

Various polyamides have been prepared in quantitative yield by carrying out the polycondensation reaction with triphenyl phosphite in NMP containing 4 wt-% LiCl (Table 10). The combination of adipic acid and diamines, especially 4,4'-diaminodiphenyl ether, yields polymers of high viscosity, whereas the use of aromatic dicarboxylic acids affords polyamides of lower viscosities, probably because of slow aminolysis of *12*.

The polycondensation reaction is more profitably utilized in the preparation of polyhydrazides and poly(amide-hydrazides) with high molecular weights[19], which are difficult to be produced by the phosphorylation reaction with triphenyl phosphite and pyridine[20]. The results of the polycondensation reaction of p-aminobenzoylhydrazide(p-ABH) with dicarboxylic acids in the presence of triphenyl phosphite and LiCl are listed in Table 11.

3.1.4 Matrix Polycondensations

Some modifications of the polycondensation in pyridine using triphenyl phosphite in the presence of PEO and P4VP give high molecular weight poly(terephthalamides) as described above. However, the polycondensation reaction with these polymer matrixes do not yield high molecular weight poly(terephthalamides) from p-phenylenediamine and 4,4'-diaminodiphenylsulfone with low basicity and poly(m-benzamide).

We have overcome the aforementioned difficulties by employing a new direct polycondensation reaction promoted by triphenyl phosphite and LiCl in NMP, espe-

Table 11. Preparation of poly (amide-hydrazides) by direct polycondensation reaction of p-ABA with dicarboxylic acids by using triphenyl phosphite and LiCl in NMP[a]

	Dicarboxylic Acid	Temp (°C)	LiCl added (g)	η_{inh}[b]
7.	HOOC—⟨C$_6$H$_4$⟩—COOH	100	1.0	0.42
			2.0	0.66
		120	0.5	1.10
			1.0	2.10
			1.5	1.78
			2.0	1.30
		140	1.0	1.40
8.	HOOC—⟨C$_6$H$_4$⟩—COOH	120	1.0	0.47
9.	HOOC–(CH$_2$)$_4$–COOH	120	1.0	0.58

[a] [p-ABH] = [Dicarboxylic Acid] = 0.01 mol; P(OPh)$_3$ = 0.02 mol; NMP = 50 ml; time = 2 h
[b] Measured at 0.5% concentration in DMSO at 30 °C

cially in the presence of poly(vinylpyrrolidone) (PVP) as a polymer matrix (Table 12)[21]. The viscosities of poly(terephthalamides) prepared from p-phenylenediamine and 4,4'-diaminodiphenylsulfone and of poly(m-benzamide) are surprisingly high, and are probably the highest values reported to date for aromatic polyamides prepared by the direct polycondensation reaction.

Table 12. Preparation of polyamides using triphenyl phosphite and LiCl in the presence of PVP[a]

Dicarboxylic Acid	Diamine	η_{inh}[b]
HOOC—⌬—COOH	H₂N—⌬—NH₂	1.12
	H₂N—⌬—CH₂—⌬—NH₂	1.55
	H₂N—⌬—O—⌬—NH₂	1.88
	H₂N—⌬—SO₂—⌬—NH₂	1.27
HOOC—⌬—COOH (m)	H₂N—⌬—NH₂	0.70
	H₂N—⌬—O—⌬—NH₂	1.25
HOOC—(CH₂)₄—COOH	H₂N—⌬—NH₂	1.55
	H₂N—⌬—CH₂—⌬—NH₂	1.30
HOOC—⌬—NH₂		1.46
HOOC—⌬—NH₂ (m)		1.22

[a] [Monomer] = 0.4 mol/l; P(OC₆H₅)₃ = 1.1 mol/mol of monomer; PVP (MW = 36 × 10⁴) = 2 g; NMP = 50 ml; LiCl = 2 g; temperature = 100 °C; time = 2 h
[b] Measured at 0.5% concentration in H_2SO_4 at 30 °C

3.2 Polypeptides

Though the preparation of polypeptides directly from free amino acids is very difficult because of their tendency to give cyclic dimers (dioxopiperazines) by ordinary methods[22], we have succeeded in obtaining linear polypeptides with relatively high molecular weights by the direct polycondensation of α-amino acids through the use of diphenyl and triaryl phosphites in pyridine. We have also obtained polypeptides with ordered sequences by the indirect polycondensation of activated derivatives of peptides, such as their active esters by ordinary methods, directly from unactivated dipeptides (Table 13)[23].

The polycondensation of amino acids was affected significantly by the solvent. Interestingly, nonpolar solvents (such as *n*-hexane) and haloalkanes (such as chloroform) gave polymers of relatively higher viscosity than highly polar aprotic solvents (such as DMF) in spite of heterogeneity of the system in these nonpolar solvents.

The triphenyl phosphite-LiCl system with PVP as a matrix accomplished in preparation of polypeptides with high molecular weight[24]. The addition of PVP significantly increased the molecular weight of polymer produced, a maximum η_{inh} value being obtained by more than *one unit mole equivalent* of PVP to monomer, although the yield decreased with increasing amount of PVP because of the difficulty in separating the polymer produced (Table 14).

The molecular weight of poly(amino acids) produced was significantly affected by the molecular weight of PVP and the addition of PVP with high molecular weight produced polymer with higher molecular weight corresponding to about one-tenth of the molecular weight of PVP (Table 15).

These results on the amount and the molecular weight of PVP suggest that the polycondensation reaction promoted by PVP takes place mainly on PVP where the monomer, triphenyl phosphite, and LiCl are absorbed and then activated *via* the phosphonium salts (*13* and *14*) like *11* and *12*. Moreover, the results also imply that the reaction may be a matrix polycondensation in which the molecular weight of polymer produced can be dependent on the molecular weight of the polymer used as a matrix.

(*13*) (*14*)

Several poly(α-amino acids) have been synthesized under the most favorable conditions, starting e.g. from L-leucine (Table 16). α-amino acids other than L-valine give polymers of high inherent viscosities in moderate yields whereas L-valine gives low η_{inh} values and low yields, probably because of lower reactivity due to steric hindrance. The inherent viscosity value, especially of poly(DL-methionine), is surprisingly high (nearly comparable to that obtained by the NCA method), and is probably the highest value reported to date for a poly(α-amino acid) prepared by the direct polycondensation reaction.

Table 13. Direct polycondensation of amino acids and dipeptides using phosphites and phosphonites in pyridine [a]

Amino acid and peptide	HO–P(OPh)$_2$ Yield (%)	η_{inh}	P(OPh)$_3$ Yield (%)	η_{inh}	Et–P(OPh)$_2$ Yield (%)	η_{inh}
Glycine	77	0.13	70	0.15	70	0.17
L-Alanine	45	0.12	100	0.14	73	0.18
L-Leucine	75	0.12	68	0.17	60	0.25
L-Phenylalanine	72	0.09	90	0.08	46	0.11
Glycylglycine	76	0.12	100	0.19	–	–
Glycyl-L-leucine	29	0.14	–	–	–	–

[a] [Monomer] = 1.25 mol/l; [HO–P(OC$_6$H$_5$)$_2$] = 1.5 mol/mol of monomer; [P(OC$_6$H$_5$)$_3$] = [C$_2$H$_5$–P(OC$_6$H$_5$)$_2$] = 1.0 mol/mol of monomer; temperature = 40 °C; time = 18 h

Table 14. Effect of the amount of PVP on the polycondensation of L-leucine by using triphenyl phosphite in NMP [a]

Amount of PVP g	unit mol/mol of monomer	Yield (%)	Polymer η_{inh} [b]	M_n [c]
0	0	72	0.14	2111
0.5	0.45	83	0.45	15272
1.0	0.90	86	0.48	17037
1.5	1.35	55	0.63	27013

[a] L-leucine = P(OPh)$_3$ = 0.01 mol; PVP (MW) = 3.6 x 10^5; LiCl = 1.0 g; NMP = 30 ml; temperature = 80 °C; time = 8 h
[b] Measured in dichloroacetic acid at 30 °C
[c] Calculated by the equation, $\eta_{inh} = 1.53 \times 10^{-3} \times M_n^{0.59}$, reported in our previous paper

Table 15. Effect of the molecular weight of PVP on the polycondensation reaction of L-leucine [a]

PVP Molecular weight	Polymer Yield (%)	η_{inh}	M_n
1 x 10^4	94	0.09	1000
4 x 10^4	86	0.14	2111
3.6 x 10^5	83	0.73	34674

[a] [L-leucine] = [P(OPh)$_3$] = 0.01 mol; PVP = 1.0 g; LiCl = 1.0 g; NMP = 30 ml; temperature = 80 °C; time = 16 h

Table 16. Polycondensation of various amino acids by means of triphenyl phosphite in the presence of PVP[a]

Amino Acid	Yield (%)	η_{inh}[b]
L-Leucine	83	0.73
DL-Leucine	75	0.62
L-Phenylalanine	66	0.65
DL-Phenylalanine	67	0.62
L-Methionine	55	0.78[c]
DL-Methionine	53	1.09[c]
L-Valine	19	0.34
β-Alanine	52	1.36[d]

[a] [monomer] = [P(OPh)$_3$] = 0.01 mol; PVP (MW = 3.6 x 10^5) = 1.0 g; LiCl = 1.0 g; solvent = NMP = 30 ml; temperature = 80 °C; time = 16 h
[b] Measured at 0.5% concentration in dichloroacetic acid at 30 °C
[c] Measured at 0.5% concentration in m-cresol at 30 °C
[d] $\eta_{sp/c}$ measured at 0.5% concentration in formic acid at 35 °C

We have applied this process to the direct polycondensation reaction of β-alanine to obtain high molecular weight poly(β-alanine) which is difficult to prepare by the NCA method, and has been produced from alternative monomers, e.g. by the ring-opening polymerization of β-propiolactam [25] or hydrogen-transfer polymerization of acrylamide [26].

The oxidation-reduction condensation with phosphites and iodine has also been applied to the polycondensation reaction of α-amino acids, but the polymer produced is of low molecular weight [27].

3.3 Polyureas and Polythioureas

The results of the preparation of polyureas under mild conditions (a pressure of less than 40 atm of carbon dioxide and a temperature around 40 °C) and of polythioureas (using diphenyl phosphite in pyridine) are compiled in Table 17 [28]. Although the preparation of polyureas from carbon dioxide under drastic conditions (high temperatures and high pressures) or from carbon oxysulfide has been reported, neither preparative method is operative under moderate conditions, nor have the methods for the synthesis of polythioureas from carbon disulfide been described.

As Table 17 shows, aromatic diamines, from which polymers with good solubility in pyridine are formed, yield higher molecular weight polymers whereas polymers from 4,4'-diaminodiphenylsulfone and p-phenylenediamine are insoluble even in HMPA and display low viscosities in sulfuric acid. On the other hand, aliphatic diamines with high basicity afford polymers of low viscosities in low yields because of the retardation of polycondensation by the formation of pyridine-insoluble and unreactive ammonium carbamate. The same also applies to the preparation of polythioureas.

Table 17. Direct polycondensation of carbon dioxide and carbon disulfide with diamines using diphenyl phosphite in pyridine[a]

Diamine	Polyurea				Polythiourea	
	Ordinary pressure		20 atm			
	Yield (%)	η_{inh}[b]	Yield (%)	η_{inh}[b]	Yield (%)	η_{inh}[d]
H$_2$N–⟨C$_6$H$_4$⟩–CH$_2$–⟨C$_6$H$_4$⟩–NH$_2$	100	0.32	100	2.24	100	0.18
H$_2$N–⟨C$_6$H$_4$⟩–O–⟨C$_6$H$_4$⟩–NH$_2$	98	0.22	100	0.51	100	0.25
[H$_2$N–⟨C$_6$H$_4$⟩–O–⟨C$_6$H$_4$⟩–]$_2$C=(CH$_3$)$_2$	95	0.44	100	1.87	78	0.15
H$_2$N–⟨C$_6$H$_4$⟩–SO$_2$–⟨C$_6$H$_4$⟩–NH$_2$	13	0.09[c]	86	0.14[c]	17	0.07
H$_2$N–⟨C$_6$H$_4$⟩–NH$_2$	100	0.08[c]	100	0.09[c]	100	0.12[c]
H$_2$NCH$_2$–⟨C$_6$H$_4$⟩–CH$_2$NH$_2$	33	0.08	46	0.13	0	—

[a] [Monomer] 0.26 Mol/l; [HO–P(OC$_6$H$_5$)$_2$] = 2.0 mol/mol of monomer; Temperature = 40 °C; time = 4 h with CO$_2$ and 6 h with CS$_2$; [b] Measured in HMPA at 30 °C; [c] Measured in H$_2$SO$_4$ at 30 °C; [d] Measured in DMSO at 30 °C

The initial pressure of carbon dioxide and the reaction temperature affect the molecular weight of the resulting polyurea, a maximum of viscosity being given by the reaction at around 40 °C at a pressure of 20 atm of carbon dioxide. Above this pressure, the viscosity decreases with pressure, dropping at 40 atm to one-fifth of that at 20 atm.

The unfavorable effects of both elevated temperature and high pressure upon the molecular weight may be caused by a depolymerization reaction or by an intermolecular or intramolecular exchange reaction between polymers, as described by Eqs. (8) and (9). These side reactions may be promoted by higher reaction temperatures and higher pressures of carbon dioxide, and also in the presence of carbon dioxide and/or diphenyl phosphite.

$$H_2N-R-NH_2 + NH_2CONH_2 \rightarrow \frac{1}{n} (\!-\!NH-R-NH-CO\!-\!)_{\overline{n}} + 2NH_3 \qquad \text{Ref. 29} \qquad (8)$$

$$R^1CONHR^2 + R^3NH_2 \rightarrow R^1CONHR^3 + R^2NH_2 \qquad \text{Ref. 30} \qquad (9)$$

3.4 Polyesters

The usual techniques of polyesterification, such as the solution[31] or interfacial[32] reaction between an aromatic acid chloride and bisphenol, are not applicable to the preparation of poly(phenyl esters) from hydroxybenzoic acids (prepared by an acid or phenyl ester exchange reaction of the acetate or phenyl ester of the acid[33]).

We have reported a convenient method for the preparation of polyesters by direct polycondensation reaction of p-hydroxybenzoic acids using triphenyl phosphite or phenylphosphine dichloride in pyridine. However, the polycondensation of m-hydroxybenzoic acid or of a combination of aromatic dicarboxylic acids and bisphenols gives polymers only in low yields[34].

Hexachlorocyclotriphosphatriazene[b] (also known as phosphonitrilic chloride trimer, PNC) has also been found to promote the polycondensation of carboxylic acids and phenols to yield carboxylic esters in high yields. This process has been successfully applied to the preparation of polyesters directly from p- and m-hydroxybenzoic acids and also from a combination of dicarboxylic acids and bisphenols (Table 18)[13], especially from 4-hydroxy-3,5-dimethoxy benzoic acid (syringic acid) in high yield.

Among the tertiary amines examined, pyridine has proved to be most effective in the polycondensation reaction, and the most favorable results have been obtained in pyridine, suggesting that this solvent is not only a simple hydrogen chloride scavenger but also involved in the reaction itself.

The amount of PNC employed affects the polycondensation, the addition of about one-third equivalent relative to one carboxy or hydroxy group resulting in high molecular weight polymers in quantitative yield.

These results suggest that three chlorine atoms from the six chlorine atoms in PNC are involved in the polycondensation reaction, activating the carboxy groups as

---

[b] 2, 2, 4, 4, 6, 6-Hexachloro-1, 3, 5, $2\lambda^5, 4\lambda^5, 6\lambda^5$-triazatriphosphorine

Table 18. Preparation of several aromatic polyesters using PNC in pyridine[a]

Bisphenol	Dicarboxylic Acid	Yield (%)	η_{inh}[b]	PMT (°C)[d]
HO–⟨C₆H₄⟩–OH	HOOC–⟨C₆H₄(m)⟩–COOH	96	0.21[c]	325
HO–⟨C₆H₄(m)⟩–OH	HOOC–⟨C₆H₄(m)⟩–COOH	88	0.25	312
[HO–⟨C₆H₄⟩–]₂C(CH₃)₂	HOOC–⟨C₆H₄⟩–COOH	100	0.34	340
HO–⟨C₆H₃(CH₃)⟩–OH	HOOC–⟨C₆H₄⟩–COOH	100	0.27	340
HO–⟨C₆H₄⟩–COOH		100	insoluble	330[e]
HO–⟨C₆H₄(m)⟩–COOH		89	0.31	198
HO–⟨C₆H₂(H₃CO)(H₃CO)⟩–COOH		100	0.80[c]	320[e]
HO–⟨C₆H₂(Cl)(Cl)⟩–COOH		100	insoluble	no peaks
H₃CO–⟨C₆H₃(OH)⟩–COOH		91	0.34	237

[a] [Monomer] = [LiCl] = 0.01 mol; PNC = 0.0035 mol; solvent = pyridine = 25 ml; temperature = = 120 °C; time = 4 h
[b] Measured at 0.5% concentration in *sym*-tetrachloroethane/phenol, 40/60 (by weight), at 30°C
[c] Measured at 0.5% concentration in p-chlorophenol at 50 °C
[d] Polymer melt temperature determined by means of the DTA
[e] High-temperature first-order transitions

has been proposed by Caglioti et al.[12], probably *via* the N-phosphonium salt of pyridine *(15)*.

$$\text{Cl}_2\text{P=N-P(Cl)}_2\text{-N=P(Cl)}_2\text{ (ring)} + \text{RCOOH} \xrightarrow{\text{Py}} \text{RCOO–P(Py}^+\text{)(=N-N-)} \quad \text{Cl}^-$$

(15)

$$\xrightarrow{\text{R'OH}} \text{RCOOR'} + \text{decomposition products}$$

A side reaction involving incorporation of the phosphorus moiety into the polymer backbone is unlikely to occur because no phosphoric acid has been detected upon oxidation of the polymer with a mixture of sulfuric acid and nitric acid.

The thermal behavior of the polymers has been evaluated by differential thermal analysis (DTA) which reveals that most of the polymers melt in the range of 310–340 °C.

On the other hand, 4-hydroxybenzoic acid and syringic acid give infusible polymers exhibiting high temperature transitions at 330 and 320 °C, respectively. The polymer obtained from 3,5-dichloro-4-hydroxybenzoic acid exhibits neither a melting point nor a transition below 400 °C.

4 References

1. Baldwin, E.: Dynamic aspects of biochemistry, p. 56. New York: Cambridge University Press 1957
2. Yamazaki, N., Higashi, F.: Bull. Chem. Soc. Japan *46*, 1235, 1239 (1973); Tetrahedron Lett. *1972*, 415; Synthesis *1974*, 436
3. Yamazaki, N., Higashi, F.: Bull. Chem. Soc. Japan *46*, 3824 (1974); *47*, 170 (1974); Synthesis *1974*, 495
4. Yamazaki, N., Higashi, F.: Tetrahedron *30*, 1323 (1974)
5. Yamazaki, N., et al.: Tetrahedron *31*, 665 (1975)
6. Yamazaki, N., Higashi, F., Iguchi, T.: Tetrahedron Lett. *1974*, 1911
7. Yamazaki, N., Higashi, F., Niwano, M.: Tetrahedron *30*, 1319 (1974)
8. Yamazaki, N., et al.: Synthesis *1979*, 355
9. Higashi, F., Goto, M., Kakinoki, H.: J. Polym. Sci., Polym. Chem. Ed. *18*, 1711 (1980)
10. Yamazaki, N., Tomioka, T., Higashi, F.: Bull. Chem. Soc. Japan *49*, 3104 (1976)
11. Grimmel, H. C.: J. Am. Chem. Soc. *68*, 529 (1946)
12. Caglioti, L., Poloni, M., Rosini, G.: J. Org. Chem. *33*, 2979 (1968)
13. Higashi, F., et al.: J. Polym. Sci., Polym. Lett. Ed.: *18*, 385 (1980)
14. Holmer, D. A., Pickett, Jr. O. A., Saunders, J. H.: J. Polym. Sci. A-1, *10*, 1547 (1972)
15. Yamazaki, N., Higashi, F.: J. Polym. Sci., Polym. Lett. Ed. *12*, 185 (1974)

16. Yamazaki, N., Matsumoto, M., Higashi, F.: J. Polym. Sci., Polym. Chem. Ed. *13*, 1373 (1975)
17. Higashi, F., et al.: J. Polym. Sci., Polym. Chem. Ed. *18*, 851 (1980)
18. Higashi, F., et al.: J. Polym. Sci., Polym. Chem. Ed. *18*, 1099 (1980)
19. Higashi, F., Kokubo, N.: J. Polym. Sci., Polym. Chem. Ed. *18*, 1639 (1980)
20. Preston, J., Hofferbert, Jr. W. L.: J. Polym. Sci., Polym. Symp. *65*, 13 (1978)
21. Higashi, F., Taguchi, Y.: J. Polym. Sci., Polym. Chem. Ed. in press
22. Bamford, C. H., Elliot, A., Hanby, W. E.: Synthetic Polypeptides, p. 63. New York: Academic Press 1956
23. Yamazaki, N., Higashi, F., Kawabata, J.: Makromol. Chem. *175*, 1825 (1974)
24. Higashi, F., Sano, K., Kakinoki, H.: J. Polym. Sci., Polym. Chem. Ed. *18*, 1841 (1980)
25. Bestian, R.: Angew. Chem. *80*, 304 (1968)
26. Breslow, D. S., Hulse, G. E., Matlack, A. S.: J. Am. Chem. Soc. *79*, 3760 (1957)
27. Guilly, L. Le, Brack, A., Sprach, G.: Makromol. Chem. *179*, 2829 (1978)
28. Yamazaki, N., Higashi, F., Iguchi, T.: J. Polym. Sci., Polym. Lett. Ed. *12*, 517 (1974)
29. Iijima, H., Asakura, M., Kimoto, K.: Kogyo Kagaku Zasshi *68*, 240 (1965)
30. Otuji, Y., Matsumura, N., Imoto, E.: Bull. Chem. Soc. Japan *41*, 1485 (1968)
31. Kantor, S. W., Holub, F. F.: US Patent 3,160,602 (1964); C. A. *62*, 5415 (1965)
32. Eareckson, W. M.: J. Polym. Sci. *40*, 399 (1959)
33. Economy, J., et al.: J. Polym. Sci., Polym. Chem. Ed. *14*, 2207 (1976)
34. Higashi, F., Kokubo, N., Goto, M.: Polym. Prepr. Japan *28*, 946 (1979)

Received May 12, 1980
T. Saegusa (editor)

Electrochemical Production of Initiators for Polymerization Processes

Giuseppe Silvestri, Salvatore Gambino, and Giuseppe Filardo

Istituto di Ingegneria Chimica, Universita di Palermo, Viale delle Scienze, Palermo, Italy

Table of Contents

I.	List of Abbreviations	28
II.	Introduction	28
III.	Electrochemical Polymerizations	28
IV.	Electrochemical Synthesis of Polymerization Catalysts	29
	1. Synthesis of Complex Catalysts	29
	a) Coordination Compounds	29
	b) Ziegler-Natta Type Catalysts	30
	c) Metallic Ions from Sacrificial Anodes	37
	2. Electrogeneration of Species Directly Reactive Towards the Monomer	40
	a) Anion Radicals	40
	b) Cation Radicals	42
	c) Free Radicals and Ions from the Solvent or the Supporting Electrolyte	44
	i) Sulfuric Acid	44
	ii) Nitrate Ion	45
	iii) Perchlorate Ion	45
	iv) Hexachloroantimonate, Hexafluorophosphate, and Tetrafluoroborate Anions	46
	v) Water	47
	vi) Alkyl- and Aryl-ammonium, -phosphonium and -arsonium Cations	47
	vii) Iodide and Triiodide Anions	48
	viii) Alkaline Metal Cations	49
	ix) Kolbe Reaction	49
V.	Conclusion	51
VI.	References	51

I. List of Abbreviations

acac	=	2,4-pentanedionate (acetylacetonate)	MeSt =	α-methylstyrene
AM	=	acrylamide	MMA =	methyl methacrylate
AN	=	acrylonitrile	NBVE =	n-butyl vinyl ether
bipy	=	bipyridine	NVCZ =	N-vinylcarbazole
Bu	=	n-butyl	OPUF =	open pore urea-formaldehyde
Cp	=	cyclopentadiene	p =	perylene
DCE	=	1,2-dichloroethane	9PA =	9-phenylanthracene
DCM	=	dichloromethane	pp =	perylene perchlorate
DEF	=	diethyl fumarate	RU =	rubrene
DMA	=	9,10-dimethylanthracene	TBAI =	tetrabutylammonium iodide
DMF	=	N,N-dimethylformamide	TBATI =	tetrabutylammonium tri-iodide
DMSO	=	dimethyl sulfoxide	THF =	tetrahydrofuran
DPA	=	9,10-diphenylanthracene	TP =	triphenylene
Et	=	ethyl	TPP =	1,3,6,8-tetraphenylpyrene
IBVE	=	isobutyl vinyl ether	TRO =	1,3,5-trioxane
MA	=	methyl acrylate	St =	styrene

II. Introduction

Catalysis enters into many of the aspects of electrochemical processes. First, in the electrocatalytic phenomena[1] taking place on the electrode surface and related to the adsorption and desorption of the substrates and to their orientation in the double layer. Second, in the homogeneous charge transfer electrocatalytic phenomena, in which the electrochemical depolarizer acts on other components of the electrolytic solution in redox reactions, and is restored to the previous oxidation state at the electrode[2]. Third, in the electrochemical synthesis of species acting as catalysts inside systems reacting chemically. In this case the amount of substrate (in moles) reacted is not in stoichiometric ratio with the circulated electrical charge. Apart from some hydrogenation reactions[3], only polymerizations belong to the last group.

III. Electrochemical Polymerizations

The different ways in which an electrochemical system can be employed to perform the polymerization of a monomer could be divided, on the basis of the mechanism, into two main fields:
   a) polymerizations in which the monomer undergoes a direct activation at the electrodes;

b) polymerizations promoted by initiators (free radicals, cations, anions, organometallic species) produced by an electrochemical reaction on depolarizers different from the monomer.

It is obvious that the distinction among these fields cannot be made until the mechanism of the electrodic events is completely understood, and the eventual role played by all the species present in the electrolytic medium is carefully investigated. In many cases some of these species have been discovered as being responsible of the initiation, but, on the contrary, sometimes a deeper insight in the process has led to the conclusion that the monomer is the direct electrodic depolarizer[4, 5].

The field taken into account in this review comprises the electrogeneration of polymerization initiators and catalysts belonging to the class (b) of the above mentioned classification, i. e. species which are not formed by direct activation of the monomers.

A further distinction, however, is necessary inside the field, as there are some differences among the electrogeneration of complex catalysts, and the direct electrodic production of inorganic or organic low molecular weight species acting as initiators.

Three excellent reviews on the electrochemical polymerizations appeared in the recent past[6, 7, 8] in which the matter is rationalized focusing the attention on the various monomers or on the reaction mechanisms. As much as possible overlapping with those reviews has been avoided here, so the reader is referred to them for the coverage of the literature before 1973.

IV. Electrochemical Synthesis of Polymerization Catalysts

1. Synthesis of Complex Catalysts

a) Coordination Compounds

The chemistry of organometallic or coordination compounds is directly related to the field of catalysis. In fact, despite the wide number of possible applications, very few examples of electrosyntheses of these compounds have been extended to their utilization as catalysts in organic processes.

In the last year some attention has been paid to the electropolymerizations in which the electrodic depolarizer is a complex between the monomer and some Lewis acids (see Ref. 7, p. 650). These researches, pioneered by Funt, were oriented towards the formation of alternating copolymers, in reactions in which the Lewis acid gives a monomer pair charge transfer complex. This research is connected to the discovery[9] that certain polar monomers, containing nitrile or carbonyl groups, (A), can complex with Lewis acids such as zinc halides. These complexes I can undergo a thermal homopolymerization by themselves, or can react with some electron donor monomers (D), giving rise to complexes like II,

$$A + ZnX_2 \longrightarrow \underset{I}{A \ldots ZnX_2} \tag{1}$$

$$A \ldots ZnX_2 + D \longrightarrow \underset{II}{D \ldots A \ldots ZnX_2} \tag{2}$$

which act as "monomers" for alternating copolymerizations.

Recently Phillips, Davies and Smith have extensively studied the copolymerization of styrene (St) and diethyl fumarate (DEF) in the presence of $ZnBr_2$ in CH_3OH both at low[10] and high[11] current densities. They made also a comparison between the electroinitiated and photoinitiated copolymerization of St–DEF with $ZnBr_2$ in the same solvent[12,13]. The most recent results are concerned with homopolymerizations in which the monomer to be activated is complexed with zinc halides in the following systems: N-vinylcarbazole with $ZnBr_2$ in acetone[14,15]; N,N'-(4,4'-sulfonyldiphenylene)dimaleimide in toluene and dimethyl sulfoxide (DMSO) with $ZnBr_2$[16]; 2-vinylnaphthalene with $ZnCl_2$ in sulfolane[17].

The oligomerizations of dienes catalyzed by nickel[18] and cobalt[19] complexes formed by electrochemical syntheses have already been described. Furthermore, it is worth mentioning here the electrosynthesis, patented by Lehmkuhl[20,21], metal chelates of which have been proposed as catalysts for some polymerization reactions. The reaction is a good example of paired electrochemical syntheses, in which both the anodic and cathodic processes contribute to the overall reaction:

$$\text{(cathode)} \quad R-OH + e^- \longrightarrow R-O^- + 1/2\, H_2 \tag{3}$$

$$\text{(anode)} \quad M \longrightarrow M^{n+} + n\, e^- \tag{4}$$

$$\text{(bulk)} \quad M^{n+} + n\, R-O^- \longrightarrow M(O-R)_n \tag{5}$$

where M = Co, Fe, Ni, Mn, Sb, Cu, Au.

The complexing agents were acetylacetonate (acac) and alcoholate anions. The electrolyses were carried in ethereal solvents, employing tetraalkylammonium salts as supporting electrolytes.

The electrochemical synthesis of metal carbonyls[22,23,24], some of which have been proposed as initiators for the polymerization of vinyl monomers (see Ref. 25, p. 436), belongs to the same field.

b) Ziegler-Natta Type Catalysts

The utilization of electrochemical syntheses for the production of Ziegler-Natta polymerization catalysts has received some consideration in view of possible industrial applications.

Quite a large quantity of information on this subject is given in the patent literature, and the need to support the patent claims has partially diverted attention away from the performance of the basic research on the systems.

A large part of the work on this matter has been done by the research group working on electroorganic chemistry at the Monsanto Research Laboratories. The electrochemical system proposed by those researchers can be schematized in the following way (DCM = dichloromethane):

(anode)   Al / DCM , $AlCl_3$, promoters / Al   (cathode)   (6)

The promoters, protic substances such as water or methanol and olefins such as cyclohexene or ethylene, are added in traces to improve the very poor conductivity of the $AlCl_3$/DCM system. The anodic dissolution of aluminium in DCM gives rise to formation of an organometal compound, the methylenebis(dichloroaluminium) $CH_2(AlCl_2)_2$, the chemical synthesis of which was already described in the literature[26].

Looking at the electrosynthesis of this haloorganoaluminium, Mottus and Ort described the system in detail[27], proposing the following sequence of electrochemical and chemical reactions:

(cathode) $CH_2Cl_2 + e^- \longrightarrow {}^\cdot CH_2Cl + Cl^-$ (7)

$CH_2Cl_2 + 2e^- \longrightarrow {:}CH_2 + 2Cl^-$ (8)

(anode) $Cl^- \longrightarrow e^- + Cl^\cdot$ (9)

$Cl^\cdot + Al \longrightarrow AlCl$ (10)

(bulk) $2CH_2{:} \longrightarrow CH_2{=}CH_2$ (11)

$CH_2{:} + CH_2Cl_2 \longrightarrow ClCH_2{-}CH_2Cl$ (12)

$AlCl + CH_2Cl_2 \longrightarrow Cl_2AlCH_2Cl$ (13)

$AlCl + Cl_2AlCH_2Cl \longrightarrow Cl_2AlCH_2AlCl_2$ (14)

The use of $(Cl_2Al)_2CH_2$ as a co-catalyst in Ziegler-Natta type polymerizations was claimed by the Monsanto research group in a patent in which no electrolytic systems were employed for the synthesis[28]. The electrosynthesis of this compound was proposed by those researchers in a series of patents in which two different approaches to the production of the catalytic system were adopted. With the first approach[29, 30] both the organoaluminium derivative and the transition metal component of the catalyst were electrolytically prepared in the same or in different cells, attending one of the following methodologies:
i) using a transition metal anode in systems analogous to system (6), and combining the electrolysis products with the methylenebis(dichloroaluminium) prepared elsewhere, to yield the desired ratio of the two components;
ii) using anodes made of strips of aluminium and strips of transition metals, electrically connected and arranged in such an electrode geometry to ensure the dissolution of both metals;

iii) using anodes made of alloys of aluminium and transition metals.

The electrolytic cell is connected to a polymerization vessel in which the electrolysis products are mixed with the olefins in the presence of a suitable solvent. The transition metals employed as anodes were vanadium and manganese. The examples quoted in the patent are related only to the polymerization of ethylene; when vanadium was employed, a polymer containing a residue of this metal not exceeding 8 p.p.m., which is a sufficiently low level for many commercial applications, was obtained.

As an example are reported the data related to a typical procedure:
a) electrolytic step: aluminium and vanadium strips as anodes, aluminium strips as cathodes. Aluminium anode area 87.7 cm^2, and vanadium anode area 15.4 cm^2. DCM 400 cm^3, water 18 µl, AlCl$_3$ 0.268 g, cyclohexene 1 cm^3. Current intensity 0.5 A, electrolysis time 40 min. Aluminium loss 0.365 g, vanadium loss 0.0092 g. Atomic ratio Al/V = 73
b) polymerization step: the DCM solution of the catalyst, as prepared in step (a), is charged to the polymerization vessel. To the stirred reactor are then added: hexane 1 l, ethylene up to $5 \cdot 10^4$ kg/m^2, hydrogen up to $5 \cdot 10^4$ kg/m^2. The pressure of ethylene is kept constant for the polymerization time (43 min);
c) isolation step: the polymer is washed with various solvents, then dried under vacuum. The solid polyethylene weighs 38.6 g and has a melt flow index of 0.0025 with a 2 kg weight, and 0.0517 with a 10 kg weight.

With the second approach to the preparation of the catalytic system, only methylenebis(dichloroaluminium) was prepared by an electrolytic reaction[31]. In the flow sheet proposed in the patent, between the electrolytic cell and the polymerization vessel a mixing reactor is interposed, where the various transition metal derivatives are added to the aluminium containing solution. Following this method other monomers, such as butadiene, 1-butene (copolymerized with ethylene), and vinyl chloride were successfully polymerized.

Various transition metal compounds were tested (cf. Table 1). In such a large number of examples a wide variety of experimental conditions is reported; the solid polymers obtained contain, after purification, not more than 20 p.p.m. of dissolved metals. For polyethylene, melting points between 125 and 134, and molecular weights between 6500 and 23000 were reported.

Two interesting features of the polymerization step are worth mentioning.

The first one is concerned with the correlation existing between they hydrogen partial pressure and the characteristics of the polyethylene obtained: the results of a series of runs are reported in Table 2. The second aspect deals with the influence exerted by low concentrations of modifiers on the catalyst activity. Table 3 summarizes the results related to this series of runs.

The technological importance of the electrosynthesis of these catalytic systems lies in the fact that it is possible to set up very easily a continuous process for the production of a cheap catalyst, which can be used as made, effluent from the electrolytic cell, without any problem related to stabilization or loss of activity by storage. Furthermore it must be added that electrochemistry affords an extremely accurate and easy way for preparing a solution of known concentration, as the quantities of the metals to be dissolved are controlled by the current imposed.

Table 1. Polymerization of olefinic compounds and Catalysts.
Polymerization runs are reported in the same patent, where Cr, Fe, V-, Mn, Ni, Co, Mo-, Zr- acetylacetonates and titanyl-acetyl-acetonate are employed. Electrolyses were carried out using CH_2Cl_2 as solvent and $AlCl_3$ as auxiliary electrolyte, between Al electrodes with concentric electrode symmetry. A teflon-coated magnetic stirring bar was provided for agitation of the electrolysis cell. Electrolyses, at constant current, were performed until the desired concentration of methylenebis(dichloroaluminium) was reached

N.	Monomer	Transition metal compounds (mmol/l)		Al/M[a] moles	$AlCl_3$/conductivity promoter moles		Catalytic solution volume l	Solid Polymer g
1	Ethylene	Cp_2TiCl_2	(20)	Al/Ti = 1	$AlCl_3/CH_3OH$	1	0.1	2.07
2	Ethylene	$CpTiCl_3$	(7)	Al/Ti = 2.1	$AlCl_3/H_2O$	1.3	0.1	5.11
3	Ethylene	Cp_2VCl_2	(1.6)	Al/V = 10	$AlCl_3/H_2O$	1	0.1	2.68
4	Ethylene	$VOCl_3$	(1.7)	Al/V = 1	$AlCl_3/H_2O$	1	0.1	2.93
5	1,3-Butadiene	TiI_4	(20)	—	$AlCl_3/H_2O$	1	0.1	4.74
6	Ethylene	Cp_2TiCl_2	(0.5)	Al/Ti = 1	—		1	57.5
7	Ethylene	$TiCl_4$	(1)	Al/Ti = 1	—		1	203.3
8	Ethylene	$TiCl_4/H_2O$[b]	(1)	Al/Ti = 1	$AlCl_3/H_2O$	1	1	53.2
9	Ethylene	n-butoxy-$TiCl_3$	(1)	Al/Ti = 1	—		1	166.9
10	Ethylene	n-butoxy-$TiCl_3$	(0.5)	Al/Ti = 1	—		1	10.6
11	Ethylene	n-butoxy-$TiCl_3$	(1)	Al/Ti = 1	—		1	93.7
12	Ethylene	n-butoxy-$TiCl_3$	(2)	Al/Ti = 1	—		1	198.5
13	Ethylene	n-butoxy-$TiCl_3$	(3)	Al/Ti = 2	—		1	90.1
14	Ethylene	n-butoxy-$TiCl_3$	(1)	Al/Ti = 0.5	—		1	10.1
15[c]	Ethylene	n-butoxy-$TiCl_3$	(2)	Al/Ti = 1	—		1	8.8
16	Ethylene/1-Butene[d]	n-dodecyloxy-$TiCl_3$	(1)	Al/Ti = 1	—		1	39.1
17	Vinyl chloride	n-dodecyloxy-$TiCl_3$	—	Al/Ti = 1	—		—	—

[a] Aluminium anodically dissolved vs. metal transition compound (molar ratio)
[b] 1 : 1 molar ratio
[c] Bu_4NCl as auxiliary electrolyte
[d] 97 : 3 molar ratio

Table 2. Ethylene polymerization. Influence of hydrogen on melt index of polyethylene in methylenebis(dichloroaluminium), $VOCl_3$ system

Run number	Amounts in mmol of MBDCA[a]	$VOCl_3$	Solvent	H_2[b] in mol %	polymer in g	polymer/$VOCl_3$ in g	I_2[c]	I_{10}[d]	I_{10}/I_2
a	1.0	0.01	CH_2Cl_2	11	45.1[e]	26 100	0.021	0.48	22.8
b	1.0	0.01	Hexane	17	37.5	21 600	0.068	1.05	15.5
c	1.0	0.01	Hexane	17	39.4	22 750	0.178	2.75	15.5
d[f]	1.0	0.01	Hexane	17	69.6	40 000	3.8	38.8	10.2

[a] MBDCA = methylenebis(dichloroaluminium)
[b] Working total pressure $5.6 \cdot 10^4$ kg/m^2
[c] I_2 = melt index of polymer product using a 2 kg weight
[d] I_{10} = melt index of polymer product using a 10 kg weight
[e] Polymer product had 2.9 p.p.m. vanadium, 5 p.p.m. chlorine and 46 p.p.m. aluminium
[f] Reactor cleaned prior to this run

Table 3. Ethylene Polymerization. Influence of modifiers on the catalyst acticity

Run number	Amounts in mmol of MBDCA[a]	$VOCl_3$	H_2O	Modifier (mmol)	Reaction time min	Reaction temp. °C	Polymer yield g	Remarks
a	0.5	0.05	0.10	$SnCl_2$ (0.05)	10	23–75	—	Solid polymer made
b	0.4	0.01	0.01		10	23–74	—	Solid polymer made
c	0.4	0.01	0.02		20	23–63	36.7	
d	0.5	0.05		$ZnCl_2$ (0.05)	10	23–70	70.6	
e	0.4	0.01		thiophene (0.01)	10	23–80	65.5	
f	0.4	0.01		thiophene (0.0233)	10	25–63	47.0	
g	1.0	0.1		triphenylphosphine (0.1)	30	24–78	74.1	

[a] MBDCA = methylenebis(dichloroaluminium)

Ercoli and coworkers investigated the same field as Monsanto, but focusing on the continuous production of the catalytic system inside the polymerization vessel.

Those researchers developed their study looking at the anodic behavior of aluminium in halogenated organic media, such as DCM or dichloroethane (DCE), containing various electrolytes. In the course of this research, the catalytic activity leading to polymerization of ethylene of the solutions obtained was tested. The current was made to flow through electrochemical systems as the following (see Table 4):

(anode)   Al / DCM or DCE, R_4NCl / stainless steel   (cathode)   (15)

Technical grade aluminium (99,0% UNI 3950) was used in these experiments[32] performed in high pressure cells[24], in which ethylene was pressurized after the circulation of variable amounts of current. DCM proved to be too reactive, giving, after the circulation of a few hundred coulombs, a violent spontaneous reaction with both the aluminium electrodes. Therefore the main part of this research has been done with DCE. Reaching, in DCE, aluminium concentrations ranging between 0,1–1 mol / l, the system acts as a good polymerization medium for getting a mixture of highly cristalline polyethylene and oils with a branched chain structure[33].

The formation of the solid polyethylene is due to the presence of transition metal impurities in the technical grade aluminium: as already observed[34] anhydrous aluminium halides react with ethylene, giving highly branched oily polymers, but not solid polyethylene. The results quoted in Table 5[35] show how the aluminium purity influences the yield in solid polymer in the converted ethylene.

Table 5. Influence of aluminium purity on the yield of solid polyethylene (PE) in DCE

Aluminium purity	PE yield
ALP 99.0 (UNI 3950)	62 %
Al 99.9 RP	41 %
ALR 99.99 (Raffinal)	0.2 %
ALR 99.99 (Zone Melted)	0 %

So, using very pure aluminium, oils can be the exclusive product of the polymerization. The mean molecular weight of these oils is ca. 500 (by cryoscopy in benzene), ratio of methyl C-H bonds to others 3:2, 20% unsaturated C-C bonds relative to the mean molecular weight of the hydrocarbon.

The electrochemical formation of the catalyst controls the kinetics of the polymerization reaction[36]. In Table 6 are shown the results of a set of experiments in which different ways of performing the polymerization are explored: in these experiments ethylene is fed to the cell after interruption of the electrolysis. In these cases the rate of conversion increases sharply with initial concentration of aluminium in solution. When ethylene (expts. 4–6) is present from the beginning of the electro-

Table 4. Continuous production of the catalytic system in ethylene polymerization

N.	Temp. °C	Anodic catalyst production		Ethylene pressure kg/m² · 10⁴		Polymer obtained		Anodic loss of Al
		Initiation circulated charge in C	Polymerization circulated charge in C	Initial	Final	solid g	oily g	g
1	0	500	660	32	15.5	8.6	5.2	0.22
2	−9	320	950	21.6	17.5	3.0	3.6	0.584
3	0	800	–	33	19.6	13.0	2.4	0.125
4	+11	900	–	36	15.5	12.1	1.0	0.231

Table 6. Polymerization of ethylene on electrolysis. (Solvent: 1,2-dichloroethane (36 g); electrolyte Bu₄NCl (0.3 g); Expts. 1–5, Al anode (99.99%, 10 cm²); expt. 6, Al anode (99.5%, 10 cm²); aluminium-net cathode. Constant current maintained in each electrolysis. Voltage in the range 30–2.5 V)

Expt.	Al dissolved g	Mean oxidation state	Electrical charge in C	Current mA	Reaction time after electrolysis h	C_4H_2 pressure kg/m² · 10⁴		Oils g	Polyethylene g
						Initial	Final		
1	0.156	1.8	1 000	25	50	21.0	17.8	1.3	—
2	0.213	1.57	1 200	25	50	21.3	16.0	2.5	—
3	0.394	1.53	2 160	25	50	23.3	0.5	10.0	—
4	0.396	1.53	2 160	50		24.0	7.3	6.3	—
5	0.380	1.59	2 160	25		23.8	9.5	5.5	—
6	0.108	1.68	650	2		28.8	26.8	0.92	0.53

lysis (conducted at constant current of 50 mA) the system reaches a stationary state where the amount of ethylene reacted (in moles) is proportional to the current density; if electrolysis is interrupted, the absorption continues with decreasing rate, however, the previous steady state is reattained if electrolysis is restarted.

The mean oxidation state of the aluminium dissolved anodically is ca. 1.5. The cathodic reaction is found to produce 1 mol Cl⁻ for each electron equivalent.

The following empirical relation

$$-d\, n_{C_2H_4}/dt = 1.3 \cdot 10^{-4} I$$

where I = current intensity
has been drawn at 0 °C, c. d.(anodic) $\leqslant 4$ mA/cm², in the steady states of the reaction, when the electrode process becomes the rate determining step of the polymerization.

To orient the reaction towards the selective formation of solid polymers, the following system has been proposed[37]:

$$\text{Al} / \text{DCE}, \text{Bu}_4\text{N Cl}, \text{C}_2\text{H}_4 / \text{M} \tag{16}$$

where M = transition metal
which was placed inside the already mentioned high pressure electrolytic cells. The current was made to flow with aluminium as anode and the transition metal as cathode until the desired quantity of Al was dissolved (equivalent to 500–1000 C), then the current was reversed and small quantities of the transition metal (equivalent to 10–60 C) were anodically dissolved in the electrolytic solution under vigorous stirring.

As summarized in Table 7, the yields in solid polyethylene (mp : initial 132 °C, final 135 °C, by differential thermal analysis (DTA); cristallinity 85–90%) are very high. The transition metals proposed for this polymerization are: Ti, V, Cr, Mn, Fe, Co, Ni, Cu. The results quoted in the patent related to the use of these metals[38] are reported in Table 8: chromium and vanadium give the highest yield in solid polyethylene, nickel gives quantitative dimerization to 1-butene, as expected from the bibliographic data related to the chemical activity of these metals.

c) Metallic Ions from Sacrificial Anodes

The electrochemical formation of catalytic systems via the anodic dissolution of metals has been studied by Koval'chuk et al.[39–42], for the polymerization of various vinyl monomers.

These authors describe the polymerization of acrylonitrile (AN), in systems in which peroxydisulfate ions $S_2O_8^{2-}$, already present in the electrolyte, are decomposed following the well known reaction[25]:

$$M^{n+} + S_2O_8^{2-} \rightarrow M^{n+1} + SO_4^{2-} + SO_4^{\cdot -} \tag{17}$$

where M is here Fe^{2+} or Cu^+

Table 7. Polymerization of ethylene with electrochemical initiator. (Solvent: 1,2-dichloroethane (36 g); electrolyte Bu$_4$NCl (0.5 g). Expts. 1–3: Al 99.99% (60 cm^2 wetted surface); electrodeposited Cr (5 cm^2); currents 5–20 mA. Expts. 4–5: technical grade Al (60 cm^2 in expt. 4; 5 cm^2 in expt. 5); current 10–2 mA (expt. 4) or 10 mA (expt. 5), during whole reaction time. Constant current maintained in each expt. Voltages in the range 30–5 V. Temperature: expts. 1,2,4, and 5: 0 °C; expt. 3: 25 °C)

Expt.	Electrical charge in C		Reaction time h	C$_2$H$_4$ pressure kg/m^2 · 10^4		Polymer obtained in g	
	Al anode	Cr anode		Initial	Final	oils	polyethylene
1	530	9	31	34.1	8.0	0.25	12.10
2	1008	18	8	29.3	1.9	0.30	16.40
3	1045	19	269	37.5	16.2	7.77	3.82
4	1160		108	32.8	16.3	5.45	8.35
5	5800		162	29.0	16.4	35.90[a]	2.30

[a] Cell refilled with C$_2$H$_4$ during reaction

Table 8. Influence of transition metal on the electrochemically initiated polymerization of ethylene

Expts.	Transition metal anode	Aluminium anodic loss [mg]	Anodic circulated charge in C		Polymerization time [h]	Converted ethylene [%]	Polymer yield [%]
			Al	Trans. metal			
1	Ti	135	900	11	62	50	69
2	V	174	850	11	64	27	95
3	Cr	112	820	10	71	91	96
4	Mn	142	950	11	72	60	—[a]
5	Fe	116	900	11	65	9.3	17
6	Ce	134	900	11	162	35	1[a]
7	Ni	107	850	5	15	99	—[b]
8	Cu	140	900	11	88	4	—[a]

[a] Oily products [b] 1-butene product

by ions such as Fe^{2+} and Cu^+ produced "in situ" by anodic dissolution of the corresponding metals. The anion radicals SO_4^- initiate the polymerization of AN. The formation of SO_4^- and the consequent polymerization are therefore controlled by the current flowing through the system.

Related to the electroinitiations with Fe^{2+}/peroxides is the paper by Mendis, Pemawansa and Hettiarachchi on the grafting of poly(methyl methacrylate) on natural rubber latex[43]. In this case Fe^{2+} ions are already in solution and, as the authors tentatively propose, MMA peroxides formed anodically could start the graft polymerization.

A systematic work on the polymerization of acrylamide (AM) at anodes of Ni, Co, Cr, Mn, Ce, in non-aqueous solvents and in the presence of $Ba(ClO_4)_2$ as supporting electrolyte, has been published by Koval'chuk[44]. The maximum conversion rate is attained at the potential at which the process of anodic dissolution of the metal takes place. Table 9 shows some of the results of this study.

Table 9. Influence of anode material on the yield and molecular weight of polymer at a current density of 40 mA/cm^2

Monomer	Anode material	Polymerization time min	Polymer yield %	Mol. weight[a]
Acrylamide	Copper	300	—	—
	Iron	300	—	—
	Nickel	60	38.40	200 000
	Cobalt	60	36.06	166 000
	Chromium	60	26.50	Insoluble
	Manganese	300	27.00	163 000
Methacrylamide	Cobalt	140	50.30	Insoluble
Acrylonitrile	Cerium	150	77.20	—

[a] not specified by the authors

This reaction seems to be specific for monomers containing amide groups (acrylamide or methacrylamide), but once these monomers are present in the electrolytic medium, other monomers, e. g., acrylonitrile, can be polymerized. The authors attribute the polymerization initiation to the electrogenerated metal ions only, but it is possible that even the perchlorate ion plays a role in the formation of the initiating species. The polyacrylamide thus obtained has an electrical conductivity 3 to 4 times higher than that of polymers obtained by the usual methods. This is due to the presence of metallic cations coordinatively bound in the polymer bulk. The presence of these cations increases the thermal stability of the polymer by 20–40%.

Koval'chuk et al. have proposed another complex system for the polymerization of acrylonitrile[45]. In this case Mn^{++} ions are oxidized on Pt anodes in aqueous solution of oxalate ions, giving rise to a MnO_2 anodic deposit. Manganese dioxide reacts with the oxalate, originating free radicals which act as polymerization initiators.

An interesting application of the technique of initiating a polymerization via the dissolution of metals is quoted in a patent[46] related to the formation of resistant images on copper by means of a "quasi-photographic" method. The structure is sandwich-like: the light activates a photoconductive layer deposited on an electrically conductive transparent support; the photoconductive layer acts as a cathode, and makes it possible to close the electrochemical circuit on a copper layer acting as anode. The copper ions, going into the intermediate layer, react with the bisulfite adduct forming free radical species which initiate the polymerization of the vinyl monomers present in the same layer. The polymers so formed remain adherent to the copper.

2. Electrogeneration of Species Directly Reactive Towards the Monomer

A large number of reactions in which a given electrogenerated intermediate, either a free radical, a cation or an anion, reacts with the monomer giving rise to the polymerization has been quoted in the literature. For many of these examples an "electrocatalytic effect" has been recognized, in which the intermediate does not form stable bonds with the monomer, but exchanges electrons returning to the previous oxidation state. The well known general scheme of this mechanism is:

$$A + e^- \longrightarrow A^{-\cdot} \tag{18}$$

$$A^{-\cdot} + S \longrightarrow A + S^{-\cdot} \tag{19}$$

where A is the species active in the electrocatalytic process, and S is the substrate, in our case the monomer.

The sequence (18) – (19) has been demonstrated to be valid even for radical cations.

In 1975 Lund and Simonet[2] have published a study on the interaction of some anionic intermediates, which give rise to an anionic electrocatalytic transfer, and various monomers. Working in very dilute solutions of the monomers they made correlations between the catalytic enhancement of the reduction peak of the intermediate and the concentration of the unsaturated substrate.

A comprehensive list of aromatic compounds which have been electrochemically reduced to anion and oxidized to cation radicals is quoted in the exhaustive review on organic electrochemistry made by Eberson and Schäfer[47].

a) Anion Radicals

The already mentioned review of Funt and Tanner[7] refers to the use until 1972 of anthraquinone, benzophenone, nitrobenzene, and naphthalene in the production of anion radicals to be employed as polymerization initiators.

Among these aromatics, only anthraquinone has subjected to further studies and will be considered in some detail here.

Anthraquinone can be reduced in two monoelectronic steps[48, 49], the first of which gives rise to the anion radical semianthraquinone, which is very stable in oxygen-free media. Semianthraquinone was proposed for the polymerization of acrylonitrile, methyl methacrylate and styrene[50, 51]; later the mechanism of the reaction with acrylonitrile has been studied by Romanin[52], who established that when oxygen is absent from the reaction medium semianthraquinone reacts with a direct electron transfer mechanism on the monomer. More recently Bezuglyi et al. have published[53] a study on the anthraquinone intermediation in the polymerization of styrene, in which, on the basis of IR spectral analysis of the obtained polymer, of polarographic data and of quantum mechanical calculations, a dimeric complex of styrene anion radical and styrene is proposed as initiator of the polymerization.

It has been demonstrated by Bhadani and Parravano that pyridine anion radicals, formed by electrochemical reduction, are able to initiate the polymerization of 4-vinylpyridine, if the monomer is added to the yellow or blue pre-electrolyzed pyridine solution[54]. The yellow color is attributed to the $Py^{-\cdot}$ and the blue one to the 4,4'-$Bipy^{-\cdot}$ anion radicals. 4-Vinylpyridine, if present during the electrolysis undergoes direct cathodic reduction giving rise to anionic polymerization.

Bhadani and Prasad[55] have recently published a paper on the cathodic polymerization of maleic anhydride in DMF, employing N-methylpyridinium iodide as supporting electrolyte. The reaction proceeds through the formation of a monomeric anion radical and subsequently a dimeric one, arising from the reduction of the N-methylpyridinium cation. These intermediates are responsible for the anionic polymerization of maleic anhydride.

Carbon dioxide has been object of detailed studies either as anion radical scavenger or as direct electrochemical substrate in organic syntheses. Among the dozens of examples of electroorganic syntheses of mono- and dicarboxylic acids even a telomerization of ethylene with CO_2 has been reported[56, 57]. The reaction has the following stoichiometry:

$$2\ CO_2 + 2\ e^- + n\ C_2H_4 \longrightarrow {}^-OOC-(CH_2CH_2)_n-COO^- \tag{20}$$

where $n = 0 - 5$

and has been realized in the following electrochemical system:

$$\text{(anode)} \quad Al\ /\ DMF\ ,\ Bu_4NBr\ ,\ CO_2,\ C_2H_4\ /\ M \quad \text{(cathode)} \tag{21}$$

The ratio among the C_n and C_{n+2} acids is constant, and the distribution curves of the products of the reaction are influenced by several factors, such as CO_2 and C_2H_4 relative concentrations in solution, temperature, and cathodic material. On the basis of the results up to now obtained, a reaction mechanism as the following has been proposed:

$$CO_2 \xrightarrow{e^-} CO_2^{-\cdot} \xrightarrow{C_2H_4} {}^\cdot CH_2CH_2-COO^- \xrightarrow{C_2H_4} {}^\cdot CH_2(CH_2)_3COO^-$$

$$\downarrow \begin{array}{c} +CO_2 \\ +e^- \end{array} \qquad \downarrow \begin{array}{c} +CO_2 \\ +e^- \end{array} \qquad \qquad \downarrow \begin{array}{c} +CO_2 \\ +e^- \end{array}$$

$$\begin{array}{ccc} COO^- & COO^- & COO^- \\ | & | & | \\ COO^- & (CH_2)_2 & (CH_2)_4 \\ & | & | \\ & COO^- & COO^- \end{array} \qquad (22)$$

b) Cation Radicals

Among the many cation radicals originated by oxidation of various aromatic compounds[47] 9,10-diphenylanthracene (DPA) and perylene have been proposed as initiators of cationic or radicalic polymerizations of various monomers.

Mengoli and Vidotto in 1971[58] proposed the use of the electrogenerated radical cation of 9,10-diphenylanthracene as initiator for cationic polymerization of styrene and n-butyl vinyl ether (NBVE).

Concerning NBVE, the authors on the basis of kinetic and spectral data, proposed the following reaction scheme:

$$DPA \longrightarrow DPA^{+\cdot} + e^-; \qquad (23)$$

$$DPA^{+\cdot} + NBVE \longrightarrow DPA \cdot NBVE^{+\cdot} \qquad (24)$$

$$DPA \cdot NBVE^{+\cdot} + DPA^{+\cdot} \longrightarrow DPA \cdot NBVE^{++} + DPA \qquad (25)$$

The reaction of $DPA^{+\cdot}$ with styrene follows a different pattern, and the authors suggested that, in this case, besides the addition reaction seen with NBVE, an electronic transfer could be operating, with a mechanism analogous to that seen for the anion radicals:

$$DPA^{+\cdot} + St \longrightarrow DPA + St^{+\cdot} \qquad (26)$$

Funt et al. studied the reaction of $DPA^{-\cdot}$ with styrene in DCM under controlled potential electrolysis[59], confirming the mechanisms proposed by Mengoli and Vidotto, and the reaction with some substituted styrene monomers[60] finding a correlation between the molecular structure of the monomers (α-methylstyrene, *trans*-β-methylstyrene, *p*-methylstyrene, and styrene) and the rate of decrease of the cation concentration.

Later the same authors[61] published a study on the reactivities of a number of electrogenerated radical cations toward St and isobutyl vinyl ether (IBVE). The reactivity comparison was made among 9-phenylanthracene (9PA), 9,10-dimethylanthracene (DMA), 9,10-diphenylanthracene (DPA), 1,3,6,8-tetraphenylpyrene (TPP), rubrene (RU), triphenylene (TP), perylene (p). The stability of the ion radicals, in the absence of the monomers, is in the order:

$TPP^{+\cdot} > Ru^{+\cdot} > DPA^{+\cdot} > p^{+\cdot} > DMA^{+\cdot} > 9PA^{+\cdot} > TP^{+\cdot}$

The reactivity towards St is higher for the inner terms of the series ($DPA^{+\cdot}$ and $p^{+\cdot}$); too stable or too reactive ions are demonstrated to be unsuitable for acting as St polymerization initiators. Mengoli and Vidotto[62] have recently published an amperometric study on the reactivity of DPA and p towards some organic monomers. In this study the authors clarified the important role played by the monomer in addressing the mechanism of the evolution of the $DPA^{+\cdot}$.

Referring to the analytical data obtained after the reaction with the monomer, the authors make a distinction among the two possibilities:

a) only 50% of the reacted radical cations are reduced to the parent neutral molecule. In this case the addition mechanism, already suggested for the reaction with BVE, takes place;

b) the radical cations are completely reduced to the parent molecule.

In this second case the reaction proceeds through an electron transfer mechanism. Monomers having nucleophilic features, such as ethyleneimine, furan, some ethers, react following scheme a). Other monomers, with no basic groups in their molecule, as styrene MeSt, Cp, undergo the electron transfer reaction.

Oberrauch et al.[63] have studied the electrochemical production and reactivity of the radical cation perylene, in the presence of various monomers. Perylene perchlorate $C_{20}H_{12}^{+\cdot} ClO_4^-$ (pp) was prepared via an electrocrystallization technique in acetonitrile or nitroethane. Styrene, α-methylstyrene, IBVE, NBVE, NVCZ, isobutene, piperylene, THF, were tested. The electronic spectra of the obtained polymers show that the polymerization takes place via electron transfer:

$$p^{+\cdot} + M \longrightarrow p + M^{+\cdot} \begin{cases} p^{+\cdot}M \longrightarrow M^+ \xrightarrow{+nM} \text{dicationic propagation} \\ + M^{+\cdot} \\ +nM \longrightarrow \text{cation radical propagation} \end{cases} \quad (27)$$

or bond formation mechanism:

$$p^{+\cdot} \xrightarrow{M} pM^{+\cdot} \xrightarrow[-p]{+p^{+\cdot}} pM^{2+} \xrightarrow{+M} pM_2^{2+} \xrightarrow{+nM} \text{dicationic propagation}$$
$$\Big\downarrow -H^+$$
$$pM^+ \xrightarrow{+nM} \text{cationic propagation} \quad (28)$$

For each monomer both the mechanisms are operating contemporarily, and, according to the authors, one may find just one or the other prevailing in the overall reaction balance.

An interesting connection with the polymerization reaction could be found with the research of Vallot and Yu[64], who published the results of a study on the electrochemical behavior of solid perylene, in aqueous media, employing a graphite-perylene powder electrode.

c) Free Radicals and Ions from the Solvent or the Supporting Electrolyte

A very large number of examples, some of which are rather ambiguous, is related with electroinitiations in which free radicals, cations or anions arise from electrochemical events where the solvent or the supporting electrolyte act, alone or together with the monomer, as electrodic depolarizer.

To add confusion to this matter, water, often present in traces, and sometimes introduced deliberately into the system (being discharged at the electrodes) is responsible for the acidity of the anolyte or the alkalinity of the catholyte. As many of the authors recognize, only electroanalytical determinations on the system composed by solvent, supporting electrolyte and monomer, associated to the usual methodologies for the individuation of the propagation mechanism, could clarify this aspect of the electropolymerizations.

It is noteworthy, in this concern, the cyclovoltammetric study of Abkulut, Fernandez and Birke[5], for the selection of unambiguous conditions for the direct electron transfer from the electrode to various monomers.

i) Sulfuric Acid

The use of aqueous sulfuric acid as a medium for the electrochemical polymerization of vinyl monomers has been object of a considerable number of investigations, focused mainly on the MMA and AN polymerization.

Recently Pistoia et al. have reinvestigated the cathodic[65] polymerization of MMA in aqueous sulfuric acid. They furnish further arguments against the thesis of the hydrogen atom initiation, which was the hypothesis made when the reaction was proposed for the first time. The initiation is explained on the basis of a participation of H_2SO_4 to the process of cathodic reduction of the MMA-peroxides[66, 67], giving rise to the alkyl-free radicals then acting as initiators.

The radical polymerization of MMA in the above conditions gives rise to polymers having too high molecular weights for having possibilities of practical applications.

Aurizi, Filippeschi and Pistoia[68] have so experienced the anodic polymerization of MMA in methanol-H_2SO_4 solutions, realizing a system in which the HSO_4^- ion is oxidized, presumably to $HSO_4^{\cdot}$ radicals, giving rise to the radical initiation of the polymerization. The authors show the dependence of both the overall MMA conversion and the viscosity average molecular weight (M_v) value of the polymer obtained, from H_2SO_4 concentration and current intensity. It is noteworthy that, with continuous current flow, almost constant M_v values are obtained as a function of time, showing a balance between the two phenomena of gel formation and increase of the formal concentration of initiating species. The use of methanol gives a substantial improvement over the use of H_2SO_4 in the electrochemical initiation of MMA polymerization. In fact, as already observed[66], in water a spontaneous polymerization takes place, so methanol allows effective electrochemical control of the concentration of the initiating species. The anodic oxidation of H_2SO_4 in water to give free radicals was proposed for the polymerization of AN[7]. A more complete picture of this polymerization, especially with regard to control of the polymerization rate and of the molecular weight, has been published recently[69].

ii) Nitrate Ion

Since the beginning of the electropolymerizations nitrate salts have not found wide application as supporting electrolytes (see Ref. [7], pp. 606–621). The latest results on the nitrate ion influence on electrochemical polymerizations are reported by Bhadani and Prasad, who have published the results of a research on the polymerization of acrylamide both in DMF and in water, employing sodium nitrate as supporting electrolyte.

In DMF[70] the polymerization takes place both in the anolyte and in the catholyte, and the authors attribute the anodic one to the species generated by oxidation of the NO_3^- ion at the anode. The polymerization is most presumably initiated by the free radical species $NO_3^{\cdot}$ or by other species arising from the evolution of $NO_3^{\cdot}$ itself. A remarkable "pre-polymerization" effect is observed: when AM is added to the anolyte obtained after electrolysis of $NaNO_3$ solution in DMF without the monomers, polymerization takes place. A definite influence of the anodic material is also observed. In water[71] only polymerization in the anolyte was observed. In this case the authors propose a reaction mechanism in which HNO_3, formed via the $NO_3^{\cdot}$ radical arising from the anodic oxidation of nitrate ion, is responsible of the initiation. Further investigations seem to be necessary for confirming this mechanism.

iii) Perchlorate Ion

Perchlorate salts are the electrolytes most widely employed in electropolymerizations. Therefore it is not surprising that in a large number of cases some interference of the ClO_4^- anion has been found in the electrochemical process.

As it shall be noted when the influence of water is examined, in many cases it is impossible to ascertain electroanalytically whether ClO_4^- is really oxidized, or the protons arising from the anodic oxidation of water are the initiators of cationic polymerizations. It must be said, in addition, that with solvents with high dielectric constant, or having basic groups in their molecule, it is meaningless to speak of $HClO_4$ formation, since molecular perchloric acid is not an actual intermediate of the process.

In these cases the presence of the anion of a strong Brönsted acid as ClO_4^- has just the influence of avoiding, due to the very low basicity, any buffer effect on the proton, allowing it to initiate the cationic polymerization.

Nevertheless in some cases the intermediate formation of the $ClO_4^{\cdot}$ radical cation effectively takes place: here the formation of $HClO_4$ comes from the $ClO_4^{\cdot}$ radicalic abstraction of a hydrogen atom from the electrolytic medium.

The following report deals with the latest results obtained when ClO_4^- salts, employed as supporting electrolytes, seem to play a definite role in the initiation process.

Cyclic Ethers. The production of commercially valuable poly(oxymethylene)s from 1,3,5-trioxane (TRO) by chemical polymerization has been object of considerable attention. The first reports on electrochemical polymerization of TRO were by Strobel and Schulz[72] who effected the electrolysis in fused monomer, and by Asahara et al.[73] who carried out the electrolysis in dry THF (see also Ref. [7], pp. 603-4).

Since then, large attention has been paid to the electrochemical homo- and copolymerization of 1,3,5-trioxane, and many of the reports quote the use of perchlorate salts as supporting electrolytes. The formation of $ClO_4^.$ radical and the subsequent production of perchloric acid, which acts as cationic initiator for the cationic polymerization of 1,3,5-trioxane has been proposed in acetonitrile[74], in benzonitrile and nitrobenzene[75, 76, 77], dichloromethane and dichloroethane[78, 79], in fused 1,3,5-trioxane[80], in 1,3,5-trioxane dispersed in hexane and heptane[81], in nitromethane[82, 83]. The polymerization of 1,3-dioxolane in DCE with Bu_4NClO_4 has been demonstrated to happen through the direct oxidation of the monomer[84]; in accord with this finding, when the electro-copolymerization of 1,3,5-trioxane with 1,3-dioxolane is realized[74, 85, 86] no direct evidence of the perchlorate oxidation is attained.

The latest reports on the THF electrochemical polymerization were published in 1973[87] and in 1974[4]. As already stated in the Introduction, the remarkable paper of Dey and Rudd[4] has given a satisfactory description of the THF initiation, in which no interference by ClO_4^- takes place.

Unsaturated Monomers. It is reported here just a brief note related to the papers in which ClO_4^- is proved or suspected to be involved in the anodic process. In some cases the only proof is given by the lowering of the pH in the anolyte, so further and deeper investigations should be necessary for a complete understanding of the mechanism. St / ClO_4^- system has been studied by Pistoia in propylene carbonate[88, 89, 90] and in dimethylsulfate[91]. In the last solvent no ClO_4^- oxidation takes place, but a significant pre-polymerization effect is reported.

The acenaphthylene/ClO_4^- system has been studied by Kikuchi and Mitoguchi in acetic anhydride and acetonitrile[92, 93] and by Kikuchi and Matsuno in nitrobenzene[94, 95] and in nitromethane[95]. The results are not conclusive with regard to the initiation mechanism, which is currently under investigation.

Cerrai, Guerra and Tricoli studied the electropolymerization of anethole in DCE employing tetraalkylammonium perchlorates as supporting electrolytes[96] evidencing the presence of $HClO_4$ in the anolyte when anethole is absent, but no conclusive evidence on the electrodic process in the presence of the monomer was achieved.

NVCZ / ClO_4^- in acetic anhydride and in acetonitrile[97, 98], MA / ClO_4^- in mixtures of water and ethylene glycol[99], indene/ClO_4^- in DCE[100] and nitrobenzene[101] have been studied in recent years. In nitrobenzene, indene is preferentially oxidized at the anode, but, adding indene after the electrolysis, prepolymerization effects, due to the oxidation of ClO_4^- were observed.

iv) Hexachloroantimonate, Hexafluorophosphate, and Tetrafluoroborate Anions

$SbCl_6^-$. Salts of the anion $SbCl_6^-$ have been employed in several examples of anodic electropolymerizations. In many cases the formation of the intermediate $SbCl_6^.$ has been inferred by the presence of $HSbCl_6$ in the anolyte, although this could not constitute a decisive proof of the hexachloroantimonate oxidation. 1,3,5-Trioxane has been polymerized with $SbCl_6^-$ salts in DCE[102, 103] and indene in DCE[100, 104] and in nitrobenzene[101]. In these anodic polymerizations the behavior of some sacrificial anodes, such as iron[101, 102, 104], nickel and copper[101, 104] has been tested.

PF_6^-. In the already quoted paper of Nakahama, Hashimoto and Yamazaki[87] a comparison was made between the electropolymerizations of MMA, St and THF, with Bu_4NClO_4 and Bu_4NPF_6 in various solvents.

BF_4^-. The anodic oxidation of BF_4^- yields BF_3; which, as has been postulated by Funt and coworkers[105, 106] and by Tidswell and Doughty[107] some years ago (see also Ref. 7 pp. 595 and 602) can initiate, in the presence of traces of protic substances, cationic polymerization processes. More recently Mengoli and Vidotto have employed Bu_4NBF_4 in the electroinitiated polymerization of TRO dispersed in n-hexane and n-heptane[81] (see page 21), getting evidence of the formation of "acid species" of boron, presumably BF_3.

v) Water

As already noted in the preceding paragraphs of this chapter, the presence of water can represent an unavoidable inconvenience when some aprotic solvents such as acetonitrile or THF, are used. Beside these reactions it is worthwhile to note here some examples of polymerizations in which the pH variations due to the discharge of water give rise to polymerizations.

The first one is related to the production of low molecular weight resols from phenol and formaldehyde[108]. The authors find strong analogies between the thermal and electrochemical production of resols, the electrochemical one being initiated by the OH^- ions produced at the cathode by water reduction.

The second example of electroinitiated polymerization deals with the formation of open pore urea-formaldehyde (OPUF) structure by electrocondensation on the anode of a prepolymerized mixture of urea-formaldehyde oligomers[109].

The reaction mechanism is still obscure, and the authors just suppose that the acidity necessary for getting the condensation of the OPUF nearby the anode, could arise from the acidic hydrolysis of Fe^{3+} cations originated by dissolution of the stainless steel anode. The applicative aspects of this system are noteworthy and further knowledge is necessary. Tidswell and Train[110, 111, 112] studied in great detail the homo- and copolymerization of vinyl acetate in aqueous emulsions. The possible initiation mechanisms are compared and the hypothesis of the formation of a "vinyl acetate-hydrogen" radical acting as true chain initiator is discussed.

vi) Alkyl- and Aryl-ammonium, -phosphonium and -arsonium Cations

Although it has been demonstrated a very long time ago[110] that quaternary cations of the V B elements are reduced through the pathway:

$$R_4N^+ + e^- \longrightarrow R^{\cdot} + R_3N \tag{29}$$

no examples of direct electroinitiation either via the $R^{\cdot}$ or via $R_3N \rightarrow$ monomer adducts have been published, except for the paper of Smith, Phillips and Davies[113], who make a comparison between the influence that the tetraalkyl and -aryl group V A halide support electrolytes exert on the electrochemically initiated polymerization of styrene. In fact this paper should be considered just as a first approach to the matter, as the data the authors quote for the different elements (N, P, As) are not consistent, all the substituents being different for the derivatives of the three elements:

$(C_2H_5)_4N^+$, $(C_6H_5)_3PCH_2C_6H_5^+$, $(C_6H_5)_4As^+$

The experimental results show a marked increase of the activity as initiator going from the nitrogen cation to the arsenic one.

One of the explanations tentatively suggested by the authors is that the three cations undergo cathodic reduction and the initiation is due to the formation of a charge transfer complex as follows

$$\underset{H}{\overset{H}{C}}=\underset{C_6H_5}{\overset{H}{C}} \quad + \quad :P\!\!\!<\!\! \longrightarrow \quad \underset{H}{\overset{H}{C}}-\underset{C_6H_5}{\overset{\ominus\,H}{C}} \quad \overset{\oplus}{\cdot}P\!\!\!<\!\! \qquad (30)$$
$$\text{I}$$

Adduct I should be responsible for the anionic polymerization, and the order of reactivity is coherent with the increasing tendency of the neutral ternary derivatives to donate electrons to the monomer giving the charge transfer complex:

As > P > N

Other possible explanations of the experimental data are proposed by the authors, taking into account the influence of the quaternary cation on the growing chain.

Collins and Thomas[114] have studied the electropolymerization of acrylonitrile, employing tetraalkylammonium and tetraphenylphosphonium salts. In both cases they found a direct activation of the monomer, and, with the phosphonium salt, they observed that electrochemical interaction takes place at the growing end of the polymer chain:

$$\sim\!\!\sim AN_n^- \; P\Phi_4^+ + e^- \longrightarrow \sim\!\!\sim AN_n^- + P\Phi_3 + \Phi\cdot \qquad (31)$$

Sato et al. studied the influence of the supporting electrolyte on the electropolymerization of acrylonitrile in DMF[115].

vii) Iodide and Triiodide Anions

Following an interesting article on the influence of electric fields on the cationic polymerization of vinyl monomers by iodine in 1,2-dichloroethane solutions[116], Giusti and coworkers published a basic research on the electrochemical activation of tetraalkylammonium iodide and triiodide in chlorinated organic solvents[117], in the absence of the monomer. The electrodic reactions are:

Cathode: $\quad I_3^- + 2e^- \longrightarrow 3I^- \qquad (32)$

$$C_2H_4Cl_2 + e^- \longrightarrow \dot{C}H_2-CH_2Cl + Cl^- \qquad (33)$$

$$\dot{C}H_2 - CH_2Cl \begin{cases} CH_2=CH_2 + Cl^- - e^- \\ CH_2=CHCl + 1/2\ H_2 \end{cases} \quad (34)$$

Anode: $\quad 3I^- \longrightarrow I_3^- + 2e^- \quad (35)$

Reactions (33), (34) take place when just I^- anion is present, or after the electrochemical consumption of I_3^- via the reaction (32). When Cl^- is present in the anolyte, authors propose, on the basis of spectrophotometric measurements, that the anodic process might be:

$$1/2\ I_2 + 2Cl^- \longrightarrow ICl_2^- + e^- \quad (36)$$

When the above mentioned processes are conducted in the presence of a monomer, such as IBVE[118], the same authors explain the polymerization taking place with TBATI with the anodic formation of iodine via the reaction:

$$2\ I_3^- \longrightarrow 3\ I_2 + 2\ e^- \quad (37)$$

while TBAI is inactive, as the anodic reaction (35) gives rise to the inactive species I_3^-. In the same article the effect of reversal of current polarity on the polymerization is discussed.

A further article of the same authors deals with the effect of a periodically reversed polarity on the electropolymerization of anethole, IBVE and cyclohexyl vinyl ether by I_2 in DCE[119].

viii) Alkaline Metal Cations

— Lithium: The anionic polymerization of some monomers, initiated by metallic lithium deposited on a cathode in organic solvents has been proposed by Albeck et al., who studied the electropolymerization of some acrylates in methanol[120, 121, 122]. This work could be correlated with that of Kikuchi and Mitoguchi, who reported the cathodic electropolymerization of acrylonitrile with $LiClO_4$ in DMF[123].
— Sodium: Many organic or inorganic sodium salts have been employed as supporting electrolyte in many electrochemical polymerizations (see Ref. 7 pp. 606-621). The latest report in which an electroinitiation seems to happen through the Na^+ discharge to metallic sodium, is published by Mengoli and Daolio[124], who studied the anionic polymerization of propylene sulfide in DMF, with Bu_4NClO_4 and $NaBF_4$. In connection with the sodium discharge at a cathode in DMF, it is worth mentioning the paper by Haas and Moreau[125], who report the chemical reaction between metallic sodium and DMF, in the presence of organic compounds containing acidic protons.

Apart from the synthetic value of the reaction, the interest of this paper lies in the fact that many side reactions involving DMF could be explained following the proposed reaction pathway.

ix) Kolbe Reaction

Since 1849 the so called "Kolbe reactions", i. e. the anodic processes involving carboxylate anions, have been employed in hundreds of electro-organic syntheses[126]. The decarboxylation of acetate ion to CH_3 radicals able to initiate the polymerization of various monomers has been proposed since the beginning of the studies on the electrochemical polymerizations (see Ref. 7 pp. 606–621). Among the latest reports in this field, that published by Albeck and Karoly[127] on the oligomerization of methyl acrylate in methanol solution is very interesting. They propose a reaction scheme as follows:

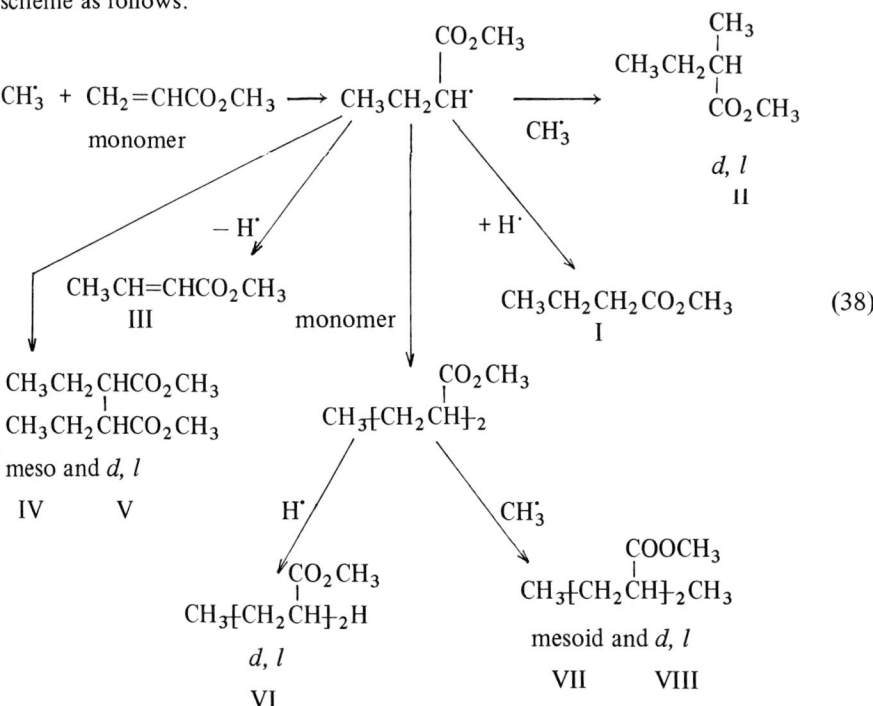

(38)

This scheme is supported by the identification of many compounds in the anolyte (from I to VIII in the scheme). The results of experiments in undivided cells are reported as well.

Several other papers have appeared dealing with the Kolbe initiation of different monomers: Tsvetkov et al.[128] studied the influence of the length of the alkyl chain of the carboxylate anion towards the acrylamide initiation; Ogumi et al.[129] published a paper on the acrylamide oligomerization initiated by anodic oxidation of trifluoroacetate anion in aqueous trifluoroacetic acid. The same authors published recently[130] a study on the current pulse electrolytic initiation of the acrylamide oligomerization in aqueous potassium acetate. The authors do not find significant differences among the products from unidirectional pulse current and steady current electrolyses. Yoshizawa et al.[131, 132] obtained high molecular weight polyacrylamide adding Fe^{III} salts to the system of the Kolbe reaction, with potassium acetate in aqueous acetic acid. Finally it is worth mentioning an interesting application of the Kolbe reaction to the telomerization of ethylene[133], in which the anodic oxidation of acetates is performed under pressures of ethylene up to 6.2×10^5 kg/m^2, obtaining variable ratios of C_6H_{14}, C_4H_{10} and C_2H_6, depending on the ethylene pressure and the composition of the liquid phase.

V. Conclusion

The most important technological interest in initiator production lies, in the electrosynthesis of complex catalytic systems, i. e. coordination compounds and Ziegler-Natta catalysts, although the choice of olefins to be polymerized with these electrosynthesized complexes is still limited to ethylene and to very few other monomers.

Looking at "in line" continuous production of initiators for the polymerization of acrylics and vinylics, the long-life cations and anions quoted here and the "pre-polymerization" effects observed in many cases make it attractive to consider these intermediates for applicative purposes as well.

Indeed, we believe that the perspectives of applicative developments of electropolymerization suffer from the same limitations one could find in other sectors of electro-organic technology: the comparison between well established "conventional" processes and processes adopting new technologies is too often favorable to the conventional ones for many and varied reasons (process experience, reliability, confidence in operation of the plants, conservative management etc.). Therefore new technologies have to prove to be much more convenient and able to solve or to overcome important and dramatic failures of the existing processes to be considered as alternatives when new industrial production is planned.

In a vast majority of cases electrochemistry just offers some moderate advantages but not revolutionary advances. Therefore a slow expansion of industrial electro-organic processes and, consequently, a problematic introduction of electropolymerization into industrial practice should be expected.

A different comment can be made on basic research: the field is very active, and combination of spectrophotometric and electroanalytical techniques for the understanding of the reaction mechanisms is usually adopted by many research groups. In fact, especially with regard to the influence of the supporting electrolyte and the evolution of the intermediate species, too often electroanalytical determinations are unadequate for the understanding of the mechanism, so the problems must be solved with still more imaginative approaches.

VI. References

1. Damaskin, B., Petrü, O, Batrakov, V.: Adsorption of Organic Compounds on Electrodes, Plenum Press, New York – London (1971)
2. Lund, H., Simonet, J.: J. Electroanal. Chem., Interfacial Electrochem. 65, 205 (1975)
3. Filardo, G. et al.: J. Chem. Soc. Dalton Trans. 1974, 1787
4. Dey, A. N., Rudd, E. J.: J. Electrochem. Soc. 121, 1294–1298 (1974)
5. Akbulut, U., Fernandez, J., Birke, R.: J. Polym. Sci., Polym. Chem. Ed. 13, 133 (1975)
6. Parravano, G., In: Baizer, M. M.: Organic Electrochemistry, New York, Dekker: 1973
7. Funt, L., Tanner, J., In: Weinberg, N. L.: Technique of electroorganic synthesis, New York – London – Sydney – Toronto: Wiley 1975
8. Shapoval, G. S., Gorodyskii, A. V., Russ. Chem. Rev. (Engl. Transl.), 42, 370 (1973)

9. Gaylord, N. G., Takahashi, A., Addition a. Condensation Polymeriz. Proc. (Platzer, N., ed.) (Adv. Chem. Ser., 91) Amer. Chem. Soc., Washington D. C. 1969, p. 94
10. Phillips, D. C., Smith, J. B., Davies, D. H.: J. Polym. Sci., Polym. Chem. Ed. *10*, 3267 (1972)
11. Phillips, D. C., Smith, J. B., Davies, D. H.: ibid., *11*, 1867 (1973)
12. Davies, D. H., Phillips, D. C., Smith, J. B.: ibid. *10*, 3253 (1972)
13. Phillips, D. C., Davies, D. H., Smith, J. B.: Makromol. Chem. *154*, 317 (1972)
14. Phillips, D. C., Davies, D. H., Smith, J. D. B.: Macromolecules *5*, 674 (1972)
15. Phillips, D. C., Davies, D. H., Smith, J. D. B.: Makromol. Chem. *169*, 177 (1973)
16. Phillips, D. C., Davies, D. H., Jackson, J. A.: Makromol. Chem. *177*, 3049 (1976)
17. Davies, D. H., Phillips, D. C., Smith, J. D. B.: J. Polym. Sci., Polym. Chem. Ed. *15*, 2673 (1977)
18. Lehmkuhl, H., In: Baizer, M. M.: Organic Electrochemistry, New York: Dekker 1973
19. Matschiner, H., Kerrinnes, H. J., Issleit, K.: Z. Anorg. Allg. Chem. *380(1)*, 1, (1971)
20. Eisenbach, W., Lehmkuhl. H., Wilke, G.: Ger. Offen. 2,349,561, Apr. 25, 1974; Chem. Abstr. *81*, 20237h (1974)
21. Lehmkuhl, H., Eisenbach, W.: Justus Liebigs Ann. Chem. *4*, 672 (1975)
22. Ercoli, R., Guainazzi, M., Silvestri, G.: J. Chem. Soc., Chem. Commun. *1967*, 927
23. Silvestri, G. et al.: J. Chem. Soc., Dalton Trans. *1972*, 2558
24. Guainazzi, M. et al.: J. Chem. Soc., Dalton Trans. *1972*, 927
25. Allen, P. E. M., Patrick, C. R.: Kinetics and Mechanisms of Polymerization Reactions, p. 433, New York: Wiley 1974
26. Lehmkuhl, H., Schäfer, R.: Tetrahedron Lett. *21*, 2315 (1966)
27. Mottus, E. H., Ort, M. O.: J. Electrochem. Soc. *117*, 885 (1970)
28. Mottus, E. H., Ort, M. O.: U. S. Pat. 3,516,978, June 23 (1970)
29. Ort, M. O. et al.: U. S. Pat. 3,546,083, Dec. (1970)
30. Mottus, E. H., Ort, M. O.: Electrochem. Soc. Meeting, Vol. 73, May 1973; Abstr. 186
31. Mottus, E. H., Ort, M. O.: U. S. Pat. 3,787,383, Jan. 22 (1974) as continuation of a series of patents on the same subject, starting with the first filing in 1966
32. Ercoli, R., Guainazzi, M., Silvestri, G.: Ger. Offen. 2,116,227 Oct. 21 (1971) as extension of Ital. pat. filed on Apr. 3, (1970)
33. Ercoli, R., Guainazzi, M., Silvestri, G.: Ger. Offen. 2,233,335, Nov. 23 (1972)
34. Thomas, C. A.: Anhydrous Aluminium Chloride in Organic Chemistry, Amer. Chem. Soc. Monogr. Ser., Reinhold Publishing Corp. p. 796 (1941)
35. Guainazzi, M., Filardo, G.: Chim. Ind. (Milan) *57*, 630 (1975)
36. Filardo, G., Guainazzi, M.: J. Chem. Soc., Dalton Trans. *1974*, 679
37. Guainazzi, M. et al.: J. Chem. Soc., Chem. Commun. *1973*, 138
38. Ercoli, R. et al.: Ger. Offen. 2,326,860, Nov. 29 (1973)
39. Koval'chuk, E. P., Tsvetkov, N. S., Fioshin, M.: Elektrokhimiya, *12*, (10) 1558 (1976)
40. Koval'chuk, E. P., Tsvetkov, N. S., Fioshin, M.: Chem. Abstr. *83*, 59418j (1975)
41. Koval'chuk, E. P., Komendat, L. N., Tsvetkov, N. S.: Chem. Abstr. *83*, 164695v (1975)
42. Koval'chuk, E. P. et al.: Elektrokhimiya *13*, (4), 614 (1977)
43. Mendis, L. P., Pemawansa, K. P. W., Hettiarachchi: Polymer *18*, 100 (1977)
44. Koval'chuk, E. P. et al.: Elektrokhimiya *13*(1), 139 (1977)
45. Koval'chuk, E. P., Tsvetkov, N. S., Gnutenko, V. A.: Chem. Abstr. *87*, 6416h (1977)
46. Levinos, S.: US Pat. 4,012,295,Mar. 15 (1977)
47. Eberson, L., Schafer, H.: Topics Curr. Chem. *21*, 1 (1971)
48. Wawzonek, S. et al.: J. Electrochem. Soc. *103*, 456 (1956)
49. Umemoto, K.: Bull. Chem. Soc. Japan *40*, 1058 (1967)
50. Trifonov, A., Panayotov, I. M.: Makromol. Chem. *77*, 237 (1964)
51. Lazarov, S., Trifonov, A., Panayotov, I.: Z. Phys. Chem. (Leipzig) *233* (1–2), 49 (1966)
52. Romanin, A. M.: Chim. Ind. (Milan) *55*, 432 (1973)
53. Bezuglyi, V. D., Shapovalov, V. A., Kovalev, I. P.: Vysokomol. Soedin., Ser. A:*18* (4), 899 (1976)
54. Bhadani, S. N., Parravano, G.: J. Polym. Sci., Part A–1, *8*, 225 (1970)
55. Bhadani, S. N., Prasad, S.: Makromol. Chem. *178*, 187 (1977)

56. Gambino, S., Silvestri, G.: Tetrahedron Lett. *32*, 3025 (1973)
57. Silvestri, G., Gambino, S., Filardo, G.: Electrochem. Soc. Spring Meet., Washington D. C. May 2–7 (1976), abstr. 278
58. Mengoli, G., Vidotto, G.: Makromol. Chem. *150*, 277 (1971)
59. Turcot, L., Glasel, A., Funt, B. L.: J. Polym. Sci., Polym. Lett. Ed. *12*, 687 (1974)
60. Funt, B. L., Verigin, V.: Can. J. Chem. *52*, 1643 (1974)
61. Glasel, A., Murray, K., Funt, B. L.: Makromol. Chem. *177*, 3345 (1976)
62. Mengoli, G., Vidotto, G.: J. Electroanal. Chem. Interfacial Electrochem. *75*, 595 (1977)
63. Oberrauch, E., Salvatori, T., Cesca, S.: J. Polym. Sci., Polym. Lett. Ed. *16* (7), 345 (1978)
64. Vallot, R., Liang, Tse Yu: C. R. Hebd. Seances Acad. Sci. Ser. C: *284*, 759 (1977)
65. Pistoia, G., Ricci, A., Voso, M. A.: J. Appl. Polym. Sci. *20*, 2441 (1976)
66. Pistoia, G., Voso, M. A.: J. Polym. Sci., Polym. Chem. Ed. *14*, 1811 (1976)
67. Schouten, A. J., Challa, G.: J. Polym. Sci., Polym. Chem. Ed. *12*, 2145 (1974)
68. Aurizi, A., Filippeschi, V., Pistoia, G.: J. Appl. Electrochem. *7*, 139 (1977)
69. Pistoia, G., Scrosati, B., Voso, M. A.: Europ. Polym. J. *12*, 53 (1976)
70. Bhadani, S. N., Prasad, Y. K.: Makromol. Chem. *178*, 1841 (1977)
71. Bhadani, S. N., Prasad, Y. K.: J. Polym. Sci., Polym. Lett. Ed. *15*, 721 (1977)
72. Strobel, W., Schulz, R. C.: Makromol. Chem. *133*, 303 (1970)
73. Asahara, T., Seno, M., Tobayama, M.: Chen Fan Quei, Seisan Kenkyu *22*, 167 (1970); Chem. Abstr. *73*, 88194g (1970)
74. Mengoli, G., Vidotto, G.: Makromol. Chem. *165*, 137 (1973)
75. Mengoli, G., Vidotto, G.: Makromol. Chem. *165*, 145 (1973)
76. Bhadani, S. N., Ghose, A.: Trans. Soc. Adv. Electrochem. Sci. Technol. *8* (1), 13 (1973)
77. Bhadani, S. N., Ghose, A.: Trans. Soc. Adv. Electrochem. Sci. Technol. *9* (4), 170 (1974)
78. Ghose, A., Bhadani, S. N.: Indian J. Technol. *12* (10), 443 (1974)
79. Mengoli, G., Valcher, S.: Europ. Polym. J. *10*, 959 (1976)
80. Mengoli, G., Vidotto, G.: Makromol. Chem. *175*, 893 (1974)
81. Mengoli, G., Vidotto, G.: J. Appl. Polym. Sci. *18*, 3095 (1974)
82. Yasuo, K., Kazuo, S.: Hiroshima Daigaku Kogakubu Kenkyu Hokoku *24*, 1 (1975); Chem. Abstr. *84*, 74660w (1975)
83. Yasuo, K., Masao, I., Kazuo, S.: ibid. *23*, 13 (1975); Chem. Abstr. *83*, 79725k (1975)
84. Mengoli, G., Valcher, S.: Europ. Polym. J. *11*, 169 (1975)
85. Fleischer, D., Schulz, R. C., Turcsanyi, B.: Makromol. Chem. *152*, 305 (1972)
86. Mengoli, G., Furlanetto, G.: Makromol. Chem. *176*, 157 (1975)
87. Nakahama, S., Hashimoto, K., Yamazaki, N.: Polym. J. *4* (4), 437 (1973)
88. Pistoia, G.: J. Polym. Sci., Polym. Lett. Ed. *10*, 787 (1972)
89. Pistoia, G.: Europ. Polym. J. *10*, 279 (1974)
90. Pistoia, G.: Europ. Polym. J. *10*, 285 (1974)
91. Pistoia, G., Scrosati, B.: Europ. Polym. J. *10*, 1115 (1974)
92. Kikuchi, Y., Mitoguchi, H.: Makromol. Chem. *173*, 233 (1973)
93. Kikuchi, Y., Mitoguchi, H.: J. Macromol. Sci., Chem. *8* (3), 573 (1974)
94. Kikuchi, Y., Matsuno, A.: Makromol. Chem. *176*, 515 (1975)
95. Kikuchi, Y., Matsuno, A.: J. Appl. Polym. Sci. *20*, 1711 (1976)
96. Cerrai, P., Guerra, G., Tricoli, M.: Europ. Polym. J. *12*, 247 (1976)
97. Kikuchi, Y., Fukunda, H.: Nippon Kagaku Kaishi *1*, 200 (1974); Chem. Abstr. *81*, 64069j (1974)
98. Kikuchi, Y., Ueyama, T.: ibid. *5*, 981 (1974); Chem. Abstr. *81*, 91991k (1974)
99. Tsvetkov, N. S., Gnutenko, V. A., Koval'chuk, E. P.: Ukr. Khim. Zh. (Russ. Ed.) *41* (4), 349 (1975)
100. Bhadani, S. N., Baranwal, P.: Makromol. Chem. *178*, 1049 (1977)
101. Bhadani, S. N., Baranwal, P.: Makromol. Chem. *178*, 2637 (1977)
102. Bhadani, S. N., Baranwal, P.: Makromol. Chem. *177*, 907 (1976)
103. Ghose, A., Bhadani, S. N.: J. Polym. Sci., Polym. Lett. Ed. *11*, 755 (1973)
104. Bhadani, S. N., Baranwal, P., Prasad, Y.: Makromol. Chem. *179*, 1623 (1978)
105. Funt, B. L., Gray, D. C.: J. Macromol. Chem. *1* (4), 625 (1966)
106. Funt, B. L., Blain, T. J.: J. Polym. Sci. Part A–1, *9*, 115 (1971)

107. Tidswell, B. M., Doughty, A. G.: Polymer *12*, 431 (1971)
108. Phillips, D. C., Meier, J. F., Jackson, J. A.: J. Polym. Sci. Polym. Chem. Ed. *12*, 1193 (1974)
109. Salyer, I. O., Usmani, A. M.: J. Appl. Polym. Sci. *22*, 3469 (1978)
110. Tidswell, B. M., Train, A. W.: Brit. Polym. J. *7*, 409 (1975)
111. Tidswell, B. M., Train, A. W.: Brit. Polym. J. *7*, 417 (1975)
112. Tidswell, B. M., Train, A. W.: Brit. Polym. J. *7*, 429 (1975)
113. Smith, J. D., Phillips, D. C., Davies, D. H.: J. Polym. Sci., Polym. Chem. Ed. *15*, 1555 (1977)
114. Collins, G. L., Thomas, N. W.: J. Polym. Sci., Polym. Lett. Ed. *13*, 73 (1975)
115. Sato, K., Ogasawara, M., Kayashi, K.: J. Polym. Sci., Polym. Lett. Ed. *11*, 5 (1973)
116. Giusti, P. et al.: Makromol. Chem. *133*, 299 (1970)
117. Cerrai, P. et al.: Europ. Polym. J. *10*, 1195 (1974)
118. Cerrai, P. et al.: Europ. Polym. J. *11*, 101 (1975)
119. Cerrai, P. et al.: Europ. Polym. J. *10*, 1141 (1974)
120. Albeck, M., Königsbuch, M., Relis, J.: J. Polym. Sci., Part A−1, *9*, 1375 (1971)
121. Albeck, M., Relis, J.: J. Polym. Sci., Part A−1, *9*, 1789 (1971)
122. Albeck, M., Relis, J.: J. Polym. Sci., Part A−1, *9*, 2963 (1971)
123. Kikuchi, Y., Mitoguchi, H.: Makromol. Chem. *175*, 687 (1974)
124. Mengoli, G., Daolio, S.: J. Polym. Sci., Polym. Lett. Ed. *13*, 743 (1975)
125. Haas, H. C., Moreau, R.: J. Polym. Sci., Polym. Chem. Ed. *16*, 699 (1978)
126. Eberson, L., In: Baizer, M. M. (ed.) Organic Electrochemistry, p. 470, New York: Dekker 1973
127. Albeck, M., Karoly, A.: J. Polym. Sci., Polym. Chem. Ed. *13*, 2699 (1975)
128. Tsvetkov, N. S., Gnutenko, V. A., Koval'chuk, E. P.: Sint. Fiz. Khim. Polim. *11*, 16 (1973)
129. Zempachi, O. et al.: Bull. Chem. Soc. Japan, *47*, 1843 (1974)
130. Zempachi, O. et al.: Bull. Chem. Soc. Japan, *49*, 2883 (1976)
131. Shiro, Y. et al.: Denki Kagaku *40*, 724 (1972)
132. Yoshizawa, S. et al.: J. Appl. Electrochem. *6* (2), 147 (1976)
133. Karapetyan, K. et al.: Elektrokhimiya *13* (4), 544 (1977)

Received July 26, 1979
W. Kern (editor)

Influence of Solvent on Free Radical Polymerization of Vinyl Compounds

Mikiharu Kamachi
Department of Polymer Science, Faculty of Science, Osaka University, Toyonaka, Osaka, Japan

Table of Contents

1 Introduction 56
2 **Historical Background** 56
   2.1 Solvent Effect on Free Radical Polymerization Rate 56
   2.2 Solvent Effect on Stereoregulation 60
3 **Solvent Effect on Free Radical Polymerization** 61
   3.1 Solvent Effect on Free Radical Polymerization Rate 62
   3.2 Elementary Rate Constants of Methacrylates 64
   3.3 Elementary Rate Constants of Vinyl Esters 66
   3.4 Termination Rate Constants 68
4 **Interpretation of the Solvent Effect on Propagation Rate Constants** . . . 69
   4.1 Polarity of Propagating Radical 70
   4.2 Chain Transfer Reaction 70
   4.3 Radical Complex Between Propagating Radical and Solvent . . . 71
   4.4 Interaction Between Monomer and Solvent 72
   4.5 Interaction Between Polymer and Solvents 74
5 **Quantum Chemical Considerations on the Radical Complex** 75
6 **Mechanistic View on Free Radical Polymerization of Vinyl Esters in Aromatic Solvents** 78
7 **Solvent Effect on Free Radical Copolymerization** 80
8 **Concluding Remarks** 83
9 **References** 84

1 Introduction

Since free radical polymerization of vinyl compounds was interpreted by a chain reaction mechanism by Staudinger, Melville, Flory, and Schulz in the middle 1930s[1], a lot of studies on the influence of solvent on free radical polymerization of vinyl monomers have been made. The following items have been proposed as probable origins for the solvent effect; (1) viscosity effect, (2) reactivity of the transferred radical, (3) modification in initiation rate, (4) formation of a radical complex between radical and solvent or monomer, (5) copolymerization with solvent. Since Bamford et al.[2] revealed influence of aromatic solvents on elementary rate constants in 1967, most studies on solvent effects in free radical polymerization have been discussed in terms of these constants. Using methyl methacrylate or styrene as a monomer[3-13], it has been demonstrated that the solvent effect is mainly due to the influence of solvent viscosity on the termination process[11-13], while a solvent effect has been observed on the propagation process. The solvent effect has been explained in terms of a donor-acceptor complex between the propagating radical and solvent[2, 3], or the viscosity effect on the propagation reaction[11].

The elementary rate constants for the polymerizations of methyl methacrylate[14], phenyl methacrylate[15], vinyl benzoate[16] and vinyl acetate[17] in various aromatic solvents have been estimated, paying particular attention to the accumulated errors involved in this estimation.

The purpose of this review article is to summarize the historical development of the solvent effect on free radical polymerization and to point out possibilities of specific interactions of the propagating radical with solvent. The effect of metal salts on the propagation process will not be described. Emphasis will be laid on the interpretation of experimental results, relating to the influence of aromatic solvents on propagation rate constants, and on the discussion for the molecular interpretation.

2 Historical Background

2.1 Solvent Effect on Free Radical Polymerization Rate

A lot of papers have been published on the effect of solvent on free radical polymerization rate. Studies on this effect have greatly been stimulated by: (1) Norrish-Trommsdorff effect, (2) Q, e-scheme in copolymerization, (3) Retardation of the polymerization rate of vinyl acetate, (4) radical complex.

Norrish et al.[18] discovered the acceleration effect in the polymerization of methyl methacrylate in 1939. In order to elucidate the cause of the effect, polymerization rates for methyl methacrylate were determined in various kinds of solvents. Schulz et al.[19, 20] reported that acceleration occurred at about 12% and 25% conversion in the bulk polymerization of methyl methacrylate at 50 and 70 °C, respectively, while there was no effect observed up to high conversion with styrene. Trommsdorff et al.[21] showed that this phenomenon is due to the increased viscosity of the polymerization system which is not caused by the interaction between a propagating radical and solvent. In order to investigate how the variation of the viscosity of the medium affects the elementary reaction process, the elementary rate constants at various conversions in different solvents were determined. The results indicate that the variation of the termination process with viscosity is much larger than that of the propagation process[22, 23].

In 1942, Alfrey and Price[24] proposed the Q, e-scheme to account for the behavior of different monomers in free radical copolymerization. They attempted to describe the reactivity of monomer by means of Q and e values according to Eqs. (2.1)

$$Q = \exp[-q/RT] \quad e = c/(rDRT)^{0.5} \tag{2.1}$$

where q = resonance energy (kcal/mol), c = induced charge, r = distance between the charge centers, and D = dielectric constant of the medium.

Since the hypothesis suggested the participation of the dielectric constant in copolymerization, the solvent effect on the reactivity ratio was investigated with several kinds of monomer pairs but could not be observed in free radical copolymerization[25–28]. Since then, it has been believed for a long time that the monomer reactivity ratio is independent of the reaction medium except for systems including ionogenic monomers such as acrylic acid and systems containing tautomers such as acrylamide. However, an appreciable effect of the reaction medium on the reactivity ratio has recently been reported[29–35]. Several factors have been proposed for the variation of the reactivity ratio as a function of the reaction medium: monomer solvation, stabilization of growing chains by complex formation and hydrogen bonding.

It was found by Burnett and Melville[36] in 1947 that the radical polymerization of vinyl acetate was retarded in aromatic solvents. This retardation effect was confirmed by several researchers[37–42]. It is characterized by three features all of which cannot be simultaneously explained by the conventional kinetic scheme involving degradative chain transfer to solvent. (1) The rate of polymerization is markedly reduced in comparison with that in many aliphatic solvents. (2) The order with respect to initiator remains close to one-half over a wide range of initiator concentration. (3) The reduction in molecular weight of the polymer is slight as compared to that in many other solvents. Stockmayer et al.[41, 42] once interpreted this retardation effect in terms of copolymerization involving the aromatic ring, but the failure of the copolymerization of benzene with vinyl monomers was confirmed by the application of the isotope technique[43–47]. Therefore, the influence of aromatic compounds on the polymerization rate of vinyl acetate has remained unsolved.

Some papers report on the influence of aromatic compounds on the polymerization of vinyl compounds other than vinyl acetate. Mayo et al.[48] found that bromobenzene acts as a chain transfer agent in the polymerization of styrene, although no fragments of bromobenzene are incorporated into the polymer. They concluded that a complex is formed between the solvent molecule and either the propagating polystyryl radical or hydrogen atom derived from the latter.

In 1957, Russell et al.[49–51] also demonstrated that the distribution of the products obtained in the photochlorination of 2,3-dimethylbutane in aliphatic solvents differed from that in aromatic solvents and that this difference was attributed to the

$$\text{Br-Ph} + \sim\sim\text{CH}_2-\overset{\bullet}{\text{CH}}-\text{Ph} \longrightarrow \left[\begin{array}{c} \sim\sim\text{CH}_2-\text{CH}-\text{Ph} \\ | \\ \text{Ph(Br)} \\ + \\ \text{Br} \\ \\ \sim\sim\text{CH}_2-\text{C}(\text{PhBr})(\text{H})- \\ + \\ \sim\sim\text{CH}=\text{CH}-\text{Ph} \end{array} \right] \xrightarrow{\text{CH}_2=\text{CH-Ph}} \begin{array}{c} \sim\sim\text{CH}=\text{CH}-\text{Ph} \\ + \\ \text{Br-Ph} \\ + \\ \text{CH}_3-\overset{\bullet}{\text{CH}}-\text{Ph} \end{array}$$

formation of a π-complex between the chlorine atom and the aromatic solvent. This complex formation was confirmed by Strong et al.[52–54] by using flash photolysis. The possibility of the complex formation was also pointed out in the addition of the trichloromethyl radical to 3-phenylpropene and to 4-phenyl-1-butene[55] or in the hydrogen abstraction of the t-butoxy radical from 2,3-dimethylbutane[56]. The concept of the radical complex was frequently used in the explanation of the unsolved problems in polymer chemistry. It was found that the bulk polymerization rate of vinyl benzoate is much lower than that of vinyl acetate in spite of their similar methyl affinities[47, 57–62]. The abnormally low polymerization rate of vinyl benzoate was explained in terms of the stabilization of the propagating radical through intramolecular reversible complex formation between the polymer radical and the aromatic ring [Eq. (2.3)] since the kinetic behavior was found to be normal, i. e. $R_p \propto [M][C]^{0.5}$ [57].

$$\sim\sim\text{CH}_2-\overset{\bullet}{\underset{H}{\text{C}}}-\text{O-CO-Ph} \rightleftharpoons \sim\sim\text{CH}_2-\text{C}(\text{O-CO-C}_6\text{H}_5\bullet)$$

Burnett et al.[63, 64] observed an anomalous rate increase in the polymerization of methyl methacrylate in halobenzene. Although the experimental data did not indicate any solvent effect on the rate of decomposition of azobisisobutyronitrile, the efficiency of initiation varied with solvents. Since an enhanced rate of incorporation of initiator fragments and incorporation of solvent fragments into the polymer were not observed, a mechanism describing the increase in the initiator efficiency through the participation of an initiator-solvent-monomer complex was postulated [Eq. (2.4)]. Henrici-Olivé et al.[65] reported, however, that the rate of the azobisisobutyronitrile decomposition at 50 °C, measured spectroscopically, is higher in halobenzene than in benzene. Burnett et al.[66] found a similar enhanced rate effect of halobenzenes with other initiators, supporting his mechanism.

Henrici-Olivé et al.[67–71] showed that this mechanism only inadequately describes the experimental data when the degree of polymerization was taken into ac-

$$\underset{\underset{CN}{|}}{CH_3-\overset{X}{\underset{|}{C}}-CH_2-H} \cdot + CH_2=\underset{\underset{COOCH_3}{|}}{\overset{CH_3}{\underset{|}{C}}}-COOCH_3 \longrightarrow \underset{}{\overset{X}{\bigcirc}} + CH_2=\underset{\underset{CN}{|}}{\overset{CH_3}{\underset{|}{C}}}-CN + CH_3-\underset{\underset{COOCH_3}{|}}{\overset{CH_3}{\underset{|}{C}}}\cdot$$

count. They proposed a reasonable explanation for the solvent effect on the polymerization rate of styrene and methyl methacrylate involving formation of a charge transfer complex between the propagating radical and the aromatic solvent. Since the typical polymer radical is expected to display the same electron affinity as tetracyanoethylene and chloranil which are known to form charge transfer complexes with aromatic solvent molecules, they considered that a polymer radical can form a complex with either a monomer or a solvent molecule, and that only the former complex leads to propagation. According to their theory, the rate of polymerization is given by Eq. (2.5).

$$-\frac{d[M]}{dt} = k_p \left(\frac{R_i}{k_t}\right)^{1/2} \frac{[M][M]_0}{[M] + [S]\tau_s/\tau_m} \tag{2.5}$$

where k_p: propagation rate constant
 k_t: termination rate constant
 R_i: initiation rate
 τ_s: lifetime of solvent-complexed radical
 τ_m: lifetime of monomer-complexed radical
 [M]: monomer concentration
 [S]: solvent concentration
 $[M]_0$: monomer concentration in bulk polymerization

Bamford and Brumby[2] measured individual rate constants in the polymerization of methyl methacrylate in several aromatic solvents, indicating that k_p and k_t varied with the type of solvents (Table 1). The k_t value is approximately inversely proportional to the viscosity of the polymerization system, while a solvent effect was observed in k_p in various aromatic solvents. They interpreted the solvent effect on k_p in terms of the complex formation proposed by Henrici-Olivé et al.[67-71].

Table 1. Rate constants for the polymerization of methyl methacrylate in different solvents at 25 °C; monomer concentration 4.69 mol/l[2]

Solvent	$k_p \times 10^{-2}$ (l/mol · sec)	$k_t \times 10^{-7}$ (l/mol · sec)	$R_i \times 10^7$ (l/mol · sec)
Benzene	2.6	2.1	0.412
Fluorobenzene	2.7	2.1	0.419
Chlorobenzene	2.8	1.95	0.422
Anisole	2.85	1.75	0.413
Bromobenzene	3.1	1.7	0.428
Benzonitrile	3.3	1.7	0.434

Bengough et al.[72] discovered that halogenated benzenes are photosensitizers in the polymerization of methyl methacrylate and other monomers and pointed out that no variation of the initiation rate with solvent is attributed to the lack of sensitivity in the viscosity method which was used to estimate the rate of initiation. Bamford and Brumby[9] reconfirmed their results and showed that the initiation reaction photosensitized through halobenzene did not occur under the experimental conditions used. The view that the formation of the radical-solvent complex was responsible for the solvent effect has later been reinforced by Burnett et al.[4-6] in the estimation of the elementary rate constants for methyl methacrylate and styrene in various aromatic solvents. A similar solvent effect was also observed in other methacrylates[11, 73, 74]. While this concept lacks direct support, indirect evidence was independently provided by Buchachenko et al.[75-77] and Burnett et al.[78] in the formation of complexes of nitroxide radical with a large number of aromatic solvents. Stability constants for the complex formation have been determined in various solvents, their variation with solvents being linearly correlated with that of the propagation rate constants for methyl methacrylate and styrene.

Schulz et al.[12, 13] estimated the elementary rate constants for methyl methacrylate in solvents whose viscosities varied by a factor of 170, indicating that the termination rate constants were inversely proportional to the viscosity of the solvents. The variation of propagation rate constants was much less than that of termination rate constants. They have also obtained similar results in the free radical polymerization of benzyl methacrylate.

Yamamoto et al.[7, 8] pointed out that variations of propagation rate constants for methyl methacrylate and styrene with solvents were too small to be distinguishable from the experimental error.

Thus, whether or not there is any interaction between the propagating radical and aromatic solvents still remains ambiguous. An accurate estimation of elementary rate constants is necessary in the first step to make clear the problem.

2.2 Solvent Effect on Stereoregulation

The influence of solvents on the microstructure of polymer was investigated in the free radical polymerization of methyl methacrylate for the first time by Fox et al. in 1962[79]. The stereochemical structure of the polymer prepared at 60 °C was, within the precision of the data, independent of the solvent used and did not measurably differ from that of the polymers obtained in bulk polymerization at the same temperature. Watanabe et al.[80] demonstrated that the temperature dependence of the microtacticity of poly(methyl methacrylate) varied with solvents and concluded that the temperature dependence was mainly caused by the polymer-solvent interaction.

Schröder et al.[81] studied the effect of solvent on the tacticity of poly(methacrylic acid). Unlike the methyl ester, the structure of poly(methacrylic acid) prepared at 60 °C was found to depend on the solvent, changing from 70% syndiotactic in xylene to 91–92% syndiotactic in polar solvents such as tetrahydrofuran and hexamethylphosphoric triamide.

Elias et al.[82-84] and Yamada et al.[87, 88] have independently reported that an isokinetic relationship exists between $(\Delta H_i^{\ddagger} - \Delta H_s^{\ddagger})$ and $(\Delta S_i^{\ddagger} - \Delta S_s^{\ddagger})$ in the free radical polymerization of methyl methacrylate and methacrylic acid in different solvents:

$$(\Delta H_i^{\ddagger} - \Delta H_s^{\ddagger}) = T_0 (\Delta S_i^{\ddagger} - \Delta S_s^{\ddagger}) + H_0 \tag{2.6}$$

where $\Delta H_i^{\ddagger}$, $\Delta H_s^{\ddagger}$, $\Delta S_i^{\ddagger}$ and $\Delta S_s^{\ddagger}$ are the activation enthalpy and entropy, respectively, for the formation of isotactic and syndiotactic diads. T_0, which is referred to as tactically isokinetic temperature, is a quantity characteristic of each monomer. They showed that the isokinetic temperature for poly(methacrylic acid) ($T_0 = 182\,°C$) is much larger than that for poly(methyl methacrylate) ($T_0 = 74\,°C$), and gave a clear interpretation why no solvent effect on the stereoregularity of poly(methyl methacrylate) was observed at 60 °C.

Yamada[87] demonstrated that the interaction between polymer and solvent affects the stereoregularity of poly(methyl methacrylate) because large enthalpy differences were observed in good solvents. In the case of poly(methacrylic acid), however, the polar effect of solvent was found to be a more important factor than the solubility of the polymer.

Based on the estimation of tacticity by NMR, an isokinetic relationship was observed for various kinds of vinyl esters by Elias et al.[85, 86] and Nozakura et al.[89]. The former pointed out a stereocontrol by hydrogen bond in the transition state as the cause of the solvent effect. Recently, Yamamoto et al.[90] investigated the solvent effect on the tacticity of poly(vinyl acetate) by ^{13}C-NMR in various solvents, indicating that there was no solvent effect on the stereoregularity.

The concept of a radical complex was also utilized in the elucidation of the high stereoregularity of poly(vinyl chloride) obtained in aromatic solvents[91, 92] and aldehydes[93]. However, there is no evidence for the formation of a radical complex in the free radical polymerization of vinyl chloride in aldehydes[94]. Elias[85] observed an isokinetic relationship in the stereoregularity of poly(vinyl chloride) prepared in different solvents.

It has been suggested that trialkylborane initiators form a complex with the radical end[95]. This has been shown by comparing the activation parameters for the tri-n-butylborane-initiated polymerization of vinyl trimethylacetate and of vinyl trifluoroacetate in various solvents with those for the AIBN-initiated polymerization (Table 2). The difference of stereoregulating activation parameters between the initiators may be due to the solvent effect on the formation of a complex of trialkylborane with the radical end.

3 Solvent Effect on Free Radical Polymerization

The effect of solvent on the rate and the elementary rate constants of free radical polymerization has been extensively studied. Because of the enormous amount of results available on the rate constants, even under the same conditions, it is difficult

Table 2. Activation parameters for the polymerization of vinyl trimethylacetate in various solvents initiated by (n-Bu)$_3$B or 2,2′-azobis-(isobutyronitrile) (AIBN)[89]

Solvent	$\Delta H_i^\ddagger - \Delta H_s^\ddagger$ cal/mol	$\Delta S_i^\ddagger - \Delta S_s^\ddagger$ cal/mol · deg.
(n-Bu)$_3$B		
Diethyl ether	430 ± 140	0.8 ± 0.6
Ethanol	400 ± 5	0.9 ± 0.1
n-Hexane	325 ± 100	0.4 ± 0.4
Bulk	270 ± 60	0.5 ± 0.2
Methanol	230 ± 15	0.1 ± 0.1
Acetone	210 ± 65	0.3 ± 0.3
AIBN		
Ethanol	710 ± 110	1.8 ± 0.4
Benzene	580 ± 30	1.2 ± 0.1
Methanol	485 ± 55	1.0 ± 0.2
n-Hexane	440 ± 90	0.7 ± 0.3
Ethyl ether	420 ± 110	0.6 ± 0.4
n-Propyl alcohol	390 ± 70	0.7 ± 0.2
Bulk	310 ± 30	0.5 ± 0.1
Acetone	−20 ± 110	−0.5 ± 0.4

to compare them with one other. In this chapter, therefore, the experimental results which were obtained by using the same instrument in our laboratory are shown and the probable origin of the solvent effect is discussed[14–17].

3.1 Solvent Effect on Free Radical Polymerization Rate

The influence of aromatic solvents on the polymerization rate for methyl methacrylate, vinyl benzoate and vinyl acetate is shown in Tables 3–6. These polymerization rates were determined in the range of monomer and initiator concentrations in which the relationship $R_p \propto [I]^{0.5}[M]^{1.0}$ holds. The optimum concentration range where this relation applies depends on the kind of monomer and decreases as follows; methyl methacrylate ≈ phenyl methacrylate > vinyl benzoate > vinyl acetate[14–17]. The pattern of variation of the polymerization rate for phenyl methacrylate with solvent is in accordance with that for methyl methacrylate and was opposite to that for vinyl acetate or vinyl benzoate: the polymerization rate for vinyl acetate in aromatic solvents is largest in benzene and smallest in benzonitrile (Table 6) while that for methyl methacrylate is largest in benzonitrile and smallest in benzene (Table 3). The highest polymerization rate of phenyl methacrylate in benzonitrile is 1.6 times as large as the lowest one in benzene, while the highest polymerization rate in vinyl benzoate is 6 times as large as the lowest one. Although the solvent effects on the polymerization rate of methacrylates are much smaller than those of vinyl esters, the variation on

Influence of Solvent

Table 3. Initiator polymerization rate, initiation rate, and derived k_p^2/k_t values for methyl methacrylate polymerization in various solvents at 30 °C[14]

Solvent	$R_p \times 10^5$ [a] (mol/l · sec)	$R_i \times 10^8$ [a,b] (mol/l · sec)	$(k_p^2/k_t) \times 10^3$ (l/mol · sec)
Anisole	1.77 ± 0.09 (3)[c]	1.09 ± 0.07 (3)[c]	7.19
Benzene	1.45 ± 0.08 (5)	1.09 ± 0.04 (4)	4.82
Fluorobenzene	1.40 ± 0.08 (3)	1.15 ± 0.05 (3)	4.26
Chlorobenzene	1.58 ± 0.05 (3)	1.08 ± 0.05 (3)	5.78
Benzonitrile	2.03 ± 0.07 (4)	1.09 ± 0.06 (3)	9.45

[a] [2,2'-azobis-(isobutyronitrile)] = 0.10 mol/l, [monomer] = 2.00 mol/l
[b] Inhibitor: 1,1'-diphenyl-2-picrylhydrazyl
[c] Number of repeated runs

Table 4. Initiator polymerization rate, initiation rate and derived k_p^2/k_t values for phenyl methacrylate polymerization in various solvents at 30 °C[15]

Solvent	$R_p \times 10^5$ [a] (mol/l · sec)	$R_i \times 10^8$ [a,b] (mol/l · sec)	$(k_p^2/k_t) \times 10^2$ (l/mol · sec)
Anisole	3.52 ± 0.03 (2)[c]	1.13 ± 0.03 (2)[c]	2.62
Benzene	3.04 ± 0.04 (4)	1.55 ± 0.05 (3)	1.49
Fluorobenzene	2.90 ± 0.03 (2)	1.40 ± 0.03 (2)	1.50
Chlorobenzene	3.40 ± 0.04 (2)	1.31 ± 0.03 (2)	2.21
Bromobenzene	3.91 ± 0.03 (2)	1.18 ± 0.03 (2)	3.24
Benzonitrile	4.36 ± 0.03 (3)	1.25 ± 0.04 (2)	3.80

[a–c] See Table 3

Table 5. Initiator polymerization rate, initiation rate, and derived k_p^2/k_t values for vinyl benzoate polymerization in various solvents at 30 °C[16]

Solvent	$R_p \times 10^7$ [a] (mol/l · sec)	$R_i \times 10^9$ [a,b] (mol/l · sec)	$(k_p^2/k_t) \times 10^5$ (l/mol · sec)
Anisole	5.66 ± 0.12 (2)[c]	6.95 ± 0.12 (2)[c]	4.52[d]
Benzene	8.15 ± 0.13 (3)	7.68 ± 0.12 (2)	8.48
Fluorobenzene	9.14 ± 0.14 (3)	7.10 ± 0.16 (3)	11.5
Chlorobenzene	7.71 ± 0.13 (2)	7.81 ± 0.14 (2)	7.46
Ethyl benzoate	3.55 ± 0.12 (2)	6.58 ± 0.15 (2)	1.90
Benzonitrile	1.77 ± 0.10 (3)	8.78 ± 0.10 (3)	0.36
Ethyl acetate	10.71 ± 0.11 (3)	7.16 ± 0.16 (2)	15.7

[a] [2,2'-azobis-(isobutyronitrile)] = 0.10 mol/l, [monomer] = 1.01 mol/l
[b] Inhibitor: p-benzoquinone
[c] Number of repeated runs
[d] Value calculated from mean values of R_p and R_i

Table 6. Initiator polymerization rate, initiation rate and derived k_p^2/k_t values for vinyl acetate polymerization in various solvents at 30 °C[17]

	$R_p \times 10^7$ [a] (mol/l · sec)	$R_i \times 10^9$ [a,b] (mol/l · sec)	$(k_p^2/k_t) \times 10^5$ (l/mol · sec)
Anisole	5.57 ± 0.51 (2)[c]	7.96 ± 0.12 (2)[c]	0.97[d]
Benzene	11.90 ± 0.90 (3)	8.23 ± 0.13 (3)	4.30
Fluorobenzene	10.21 ± 0.80 (2)	8.61 ± 0.15 (2)	3.02
Chlorobenzene	7.02 ± 0.24 (2)	8.84 ± 0.12 (2)	1.39
Ethyl benzoate	3.45 ± 0.26 (2)	8.90 ± 0.13 (2)	0.33
Benzonitrile	0.92 ± 0.018 (2)	8.99 ± 0.14 (2)	0.024
Ethyl acetate	123 ± 12	8.78 ± 0.10 (2)	430.8

[a] [2,2'-azobis-(isobutyronitrile)] = 0.10 mol/l, (Monomer) = 2.00 mol/l
[b–d] See Table 5

methacrylates is still worthy of further consideration, because the polymerization rate was dilatometrically measured with the experimental error of 3%. Since the anomaly in the polymerization of methyl methacrylate in halobenzene has been interpreted as the participation of the solvent in the initiation process[63, 64, 66], the solvent effect on the initiation rate was first investigated. Estimation of the initiation rate was performed by the inhibition method using DPPH and benzoquinone as inhibitor in the polymerization of methacrylates and vinyl esters, respectively. The results are shown in Tables 3–6. It can be seen that the initiation rates are slightly affected by the solvents and that the observed variation in the polymerization rate mainly comes from variation of the propagation or termination process. An anomaly in halobenzenes could not be found under experimental conditions, which is consistent with the results of Bamford et al.[2]. In order to make clear the origin of the solvent effect, the influence of solvent on elementary rate constants was investigated.

3.2 Elementary Rate Constants for Methacrylates

The elementary rate constants were calculated from ratio k_p^2/k_t, obtained from the polymerization rate and initiation rate and the ratio k_p/k_t, estimated from the lifetime of the radical determined by the rotating sector method. The mean lifetime of the propagating radical and derived rate constants for methacrylates are shown in Tables 7–8. The variation of the propagation rate constant for methyl methacrylate with solvents is in accordance with the result obtained by Bamford et al.[2] at 25 °C. Since the largest and the smallest k_p value for phenyl methacrylate differ by a factor of 1.6 and for methyl methacrylate by a factor of 1.4, the estimation of the rate constants must be performed under experimental conditions in which the accumulated error is so small as to permit a distinction of the difference. Therefore, particular attention was given to the constancy of the reaction temperature (±0.001 °C), constancy of light source, purity of monomers and solvents, and reproducibility of observed values and to the retention of the square wave in the rotat-

Influence of Solvent

Table 7. Mean lifetime of the propagating radical of phenyl methacrylate and derived rate constants in various solvents at 30 °C[15]

	τ^a (sec)	$(k_p/k_t) \times 10^4$	$k_p \times 10^{-2}$ (l/mol · sec)	$k_t \times 10^{-6}$ (l/mol · sec)
Anisole	1.21 ± 0.08 (4)[b]	1.14	2.30	2.02
Benzene	1.00 ± 0.04 (8)	0.85	1.76	2.06
Fluorobenzene	1.13 ± 0.06 (6)	0.83	1.80	2.16
Chlorobenzene	1.15 ± 0.07 (6)	0.99	2.23	2.26
Bromobenzene	1.21 ± 0.04 (4)	1.38	2.35	1.72
Benzonitrile	1.18 ± 0.05 (4)	1.40	2.73	1.96

[a] [2,2'-Azobis-(cyclohexane-1-carbonitrile)] = 3.00 × 10⁻³ mol/l, [monomer] = 2.00 mol/l
[b] Number of data for the determination of lifetime

Table 8. Mean lifetime of the propagating radical of methyl methacrylate and derived rate constants in various solvents at 30° [14]

	τ^a (sec)	$(k_p/k_t) \times 10^5$	$k_p \times 10^{-2}$ (l/mol · sec)	$k_t \times 10^{-7}$ (l/mol · sec)
Anisole	0.46 ± 0.04 (3)[b]	1.42[c]	5.06	3.59
Benzene	0.42 ± 0.03 (4)	1.07	4.50	4.20
Fluorobenzene	0.41 ± 0.05 (3)	0.95	4.48	4.72
Chlorobenzene	0.43 ± 0.04 (3)	1.16	4.98	4.29
Benzonitrile	0.45 ± 0.03 (3)	1.54	6.14	3.99

[a, b] See Table 7
[c] Values calculated from mean values of τ

ing sector. The limit for the accumulated error[15] was estimated by the following Eqs.,

$$\left|\frac{\Delta k_p}{k_p}\right| = \left|\frac{\Delta R_p}{R_p}\right| + \left|\frac{\Delta R_i}{R_i}\right| + \left|\frac{\Delta \tau}{\tau}\right| + \left|\frac{\Delta W_m}{W_m}\right| + \left|\frac{\Delta W_c}{W_c}\right| \qquad (3.2.1)$$

$$\left|\frac{\Delta k_t}{k_t}\right| = \left|\frac{\Delta R_i}{R_i}\right| + 2\left|\frac{\Delta \tau}{\tau}\right| + \left|\frac{\Delta W_c}{W_c}\right| \qquad (3.2.2)$$

where W_m and W_c are the weights of monomer and initiator, respectively. The limit for the error in k_p and k_t was estimated from the mean deviation of the rate and the lifetime to be around 15% and 25% for phenyl methacrylate and methyl methacrylate, respectively. Accordingly, the variation of k_p or k_t with solvent is beyond the limit of error, being worthy of further considerations on the origin of the solvent effect.

In order to show the dependence of k_p on the substituent of aromatic solvents, the relationship between the k_p of all monomers and Hammett's constant σ_p is plot-

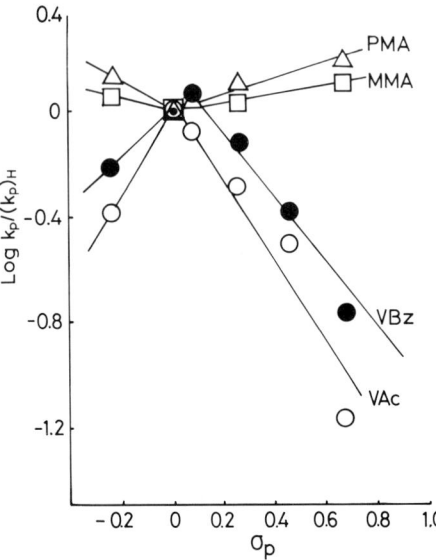

Fig. 1. Hammett's plot of solvent effect on k_p for vinyl compounds

ted in Fig. 1. The k_p values for methacrylate are shifted to higher values with substituents of positive and negative σ_p. The extent of variation of k_p with solvent is larger in the free radical polymerization of phenyl methacrylate than in that of methyl methacrylate. Henrici-Olivé and Olivé[67–71] pointed out that the electron affinities of the polymer radical are of the same order as those of molecules such as tetracyanoethylene and chloranil which are known to form charge transfer complexes with typical aromatic solvents. If the polymethacrylate radical is an electron acceptor with respect to solvent, the poly(phenyl methacrylate) radical will be a slightly stronger acceptor than the poly(methyl methacrylate) radical because the phenyl group decreases the electron density of the propagating radical as compared to the methyl group. Accordingly, the poly(phenyl methacrylate) radical is more likely to form a complex.

The k_p value for phenyl methacrylate (Table 7) is smaller than that for methyl methacrylate (Table 8), although phenyl methacrylate is more likely to be attacked by a free radical than methyl methacrylate (see copolymerization data[96,97]). Accordingly, it is clear that the propagating radical of methyl methacrylate is more reactive than that of phenyl methacrylate. This is because the phenyl methacrylate radical is more likely to form the radical-solvent complex, which is consistent with the above-mentioned proposal by Henrici-Olivé et al.[67–71] and Bamford et al.[2].

3.3 Elementary Rate Constants for Vinyl Esters

Mean lifetimes and derived elementary rate constants for vinyl esters are shown in Tables 9–10. For vinyl benzoate, there is a sevenfold difference between the largest and smallest k_p values in aromatic solvents. The k_p value in ethyl acetate is larger than that in the aromatic solvents. A similar but more drastic solvent effect was ob-

Influence of Solvent

Table 9. Mean lifetime of the propagating radical of vinyl benzoate and derived rate constants in various solvents at 30 °C[16])

	τ^a (sec)	$(k_p/k_t) \times 10^6$	$k_p \times 10^{-2}$ (l/mol · sec)	$k_t \times 10^{-7}$ (l/mol · sec)
Anisole	0.14 ± 0.03 (4)[b]	4.40[c]	1.06	2.53
Benzene	0.11 ± 0.03 (8)	4.61	1.85	4.02
Fluorobenzene	0.11 ± 0.03 (4)	4.71	2.45	5.22
Chlorobenzene	0.13 ± 0.02 (4)	4.48	1.68	3.83
Ethyl benzoate	0.11 ± 0.03 (4)	2.38	0.69	2.57
Benzonitrile	0.11 ± 0.03 (4)	1.10	0.33	2.99
Ethyl acetate	0.10 ± 0.03 (7)	5.93	2.67	4.51

[a] [2,2'-Azobis-(cyclohexane-1-carbonitrile)] = 3,00 x 10^{-3} mol/l, [monomer] = 1.01 mol/l
[b] Number of data for the determination of lifetime
[c] Values calculated from mean values of τ

Table 10. Mean lifetime of the propagating radical of vinyl acetate and derived rate constants in various solvents at 30 °C[17])

Solvent	τ^a (sec)	$(k_p/k_t) \times 10^7$	$k_p \times 10^{-2}$ (l/mol · sec)	$k_t \times 10^{-8}$ (l/mol · sec)
Anisole	0.061 ± 0.011 (4)[b]	2.01[c]	0.48	2.39
Benzene	0.065 ± 0.012 (8)	3.67	1.17	3.19
Fluorobenzene	0.066 ± 0.012 (4)	3.12	0.97	3.11
Chlorobenzene	0.065 ± 0.018 (4)	2.29	0.61	2.66
Ethyl benzoate	0.065 ± 0.016 (4)	0.897	0.37	4.12
Benzonitrile	0.059 ± 0.017 (4)	0.310	0.08	2.58
Ethyl acetate	0.170 ± 0.042 (2)	67.6	6.37	0.94
Ethyl acetate[d]	0.086 ± 0.016 (2)	6.85	1.59	2.32

[a] [2,2'-Azobis-(cyclohexane-1-carbonitrile)] = 3.03 x 10^{-3} mol/l, [monomer] = 2.00 mol/l
[b,c] See Table 9
[d] Ethyl acetate contains 2.00 mol/l of ethyl benzoate

served for vinyl acetate as shown in Fig. 1. For example, k_p in benzene is fifteen times as large as the smallest k_p in benzonitrile, and that in ethyl acetate is eighty times as large as the smallest one.

The k_p values for vinyl esters are shifted to lower values in aromatic solvents with either donor or acceptor substituents. The variation of k_p with solvent is larger for vinyl acetate than that in the case of vinyl benzoate. The variation of k_p for vinyl esters with solvent is much larger than that for methacrylates. The k_p value for vinyl benzoate is larger than that for vinyl acetate. Since vinyl benzoate contains an aromatic ring, the influence of the latter must be considered in the comparison of k_p in ethyl acetate between vinyl acetate and vinyl benzoate. Because the participation of the aromatic ring of vinyl benzoate in the propagation process was considered to be close to that of the aromatic ring of ethyl benzoate, k_p for vinyl acetate was determined to be 159 l/mol · sec in ethyl acetate containing the same concentration of ethyl benzoate as that of vinyl benzoate in ethyl acetate. Thus, the observed k_p for vinyl acetate is always smaller than that for vinyl benzoate in all solvents examined. If the propagating radical of vinyl benzoate is stabilized by intramolecular complexation[47], k_p for vinyl benzoate would have to be smaller than that for vinyl acetate, because both monomers are known to display similar reactivity toward the methyl radical[47] (methyl affinity: 36 for vinyl benzoate and 42 for vinyl acetate). This is however not the case. Therefore, the propagating radical of vinyl benzoate is not considered to be more stabilized than that of vinyl acetate, the intramolecular complex mechanism being ruled out owing to the low rate of bulk polymerization of vinyl benzoate[121].

3.4 Termination Rate Constants

North et al.[98–101] noted experimentally and theoretically that the termination rate constant in radical polymerization is inversely proportional to viscosity of the polymerizing system. Yokota et al.[102] reconfirmed this relation and pointed out

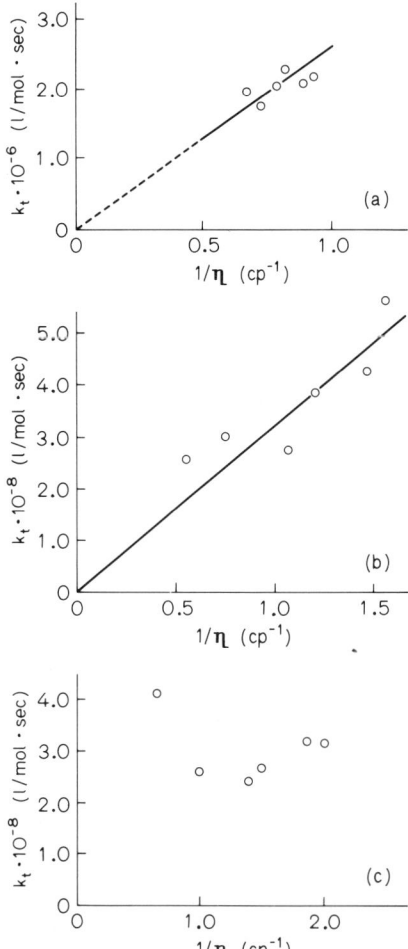

Fig. 2a–c. Correlation of k_t values with viscosity of the polymerization system.
(a) Phenyl methacrylate, (b) vinyl benzoate, (c) vinyl acetate

that the viscosity also affected the kinetic order of monomer in the polymerization rate. Bamford et al.[2] determined k_t for methyl methacrylate in aromatic solvents, indicating that his results are in excellent agreement with the relation $k_t = 1/\eta$. Fischer et al.[12, 13] determined the elementary rate constants for methyl methacrylate in solvents whose viscosities varied by a factor of 170. The rate of chain termination was confirmed to be inversely proportional to the viscosity of the polymerization medium.

In the estimation of k_t, the correlation between k_t and the reciprocal of the viscosity of the polymerization system was investigated for all monomers in Fig. 2. The k_t values for phenyl methacrylate seem to be distributed near the straight line through the origin (Fig. 2a), being consistent with Bamford's result[2]. The k_t values for vinyl benzoate seem to be distributed around the straight line (Fig. 2b), but the extent of scatter in k_t for vinyl benzoate is larger than that for phenyl methacrylate. The k_t value for vinyl acetate was not correlated with the fluidities of the polymerization system. Recently, a similar effect on k_t for vinyl acetate was independently reported by Yamamoto et al.[103]. The observed effect suggests that the termination reaction is controlled by some factor in addition to the diffusion of the polymer end.

When it is considered that the reversible complex of the propagating radical is most likely to form with vinyl acetate, a correlation of k_t with the calculated delocalization stabilization for the complex formation (see Chap. 5) will be expected in the system. However, a linear correlation could not be found. Since the activation energy for the termination is much smaller than that for the propagation reaction, termination is assumed to be less influenced by the complex formation than propagation. However, the deviation from the linear relationship in Fig. 2 becomes more pronounced in the case of monomers whose propagating radicals interact more strongly with aromatic solvents. Accordingly, the departure is possibly ascribed to the complex formation between the propagating radical and solvent.

4 Interpretation of Solvent Effect on Propagation Rate Constants

The solvent effect on free radical polymerization is presumed to be due to the following factors:
1) Polarity of the propagating radical,
2) the difference in chain transfer reaction and/or reinitiation from the solvent radical,
3) monomer solvation,
4) polymer solvation,
5) formation of a radical complex between the propagating radical and solvent,
6) copolymerization.

Since copolymerization of vinyl compounds with aromatic solvents[43–47] was confirmed by isotope technique to be absent, copolymerization was not considered in the following discussion.

4.1 Polarity of Propagating Radical

The propagating radical of vinyl esters has been reported to be electron-rich[104], while those of methacrylates are thought to be electron-poor[105]. Moreover, the inverse solvent effect (Fig. 1) seems to support the polar nature of the propagating radical. A correlation would be expected to exist between k_p and $1/\epsilon$ or $(\epsilon - 1)/(2\epsilon + 1)$ of the solvent. However not any correlation was found for all monomers used. Accordingly, the influence of solvent polarity on k_p seems to be ruled out. Moreover, a remarkable decrease of k_p for vinyl esters in ethyl benzoate as compared to that in ethyl acetate in spite of same dielectric constants of the two solvents ($\epsilon = 6.02$[106]) supports the above conclusion.

4.2 Chain Transfer Reaction

Another possible explanation for the solvent effect might be based on the difference in the chain transfer rate from the propagating radical to solvent and on the reinitiation rate by the resulting solvent radical. Let us discuss the effect of the solvent transfer reaction on k_p under the following three aspects: (1) likelihood of chain transfer to solvent, (2) stability of the resulting solvent radical, (3) polarity of the radical.

Likelihood of chain transfer reaction to solvent is much larger in the free radical polymerization of vinyl esters than in that of methacrylates, since the constants of chain transfer to solvent for the poly(vinyl ester) radical are much larger than those for the polymethacrylate radical[107]. Thus, the fact that solvent effects on k_p for vinyl esters are clearly more pronounced than those for methacrylates seems to reflect the difference in chain transfer reaction to solvent. However, the data on chain transfer for vinyl acetate[107] show that the solvent dependence of k_p does not correlate with the chain transfer to solvent. For vinyl benzoate, only two constants of chain transfer to aromatic solvents are available, the constant obtained with benzene (7.5×10^{-4}) being larger than that obtained with isopropyl benzoate (1×10^{-4}) as a solvent[57]. Since the constants for chain transfer to ethyl benzoate are considered to be smaller than that to isopropyl benzoate, chain transfer to benzene must be larger than that to ethyl benzoate. If chain transfer to solvent reduces the observed k_p values, these should become smaller in benzene than those in ethyl benzoate. This expectation is also inconsistent with the case considered. Moreover, since no isotope effect on k_p was observed in deuterated benzene for both vinyl acetate and methyl methacrylate (Table 11), it is unambiguously concluded that chain transfer reaction does not participate in the solvent effect on k_p.

The solvent effect on k_p might be due to the variation of the reinitiation rate of the solvent radical produced by solvent transfer reaction. If the stability of the solvent radical is important for the solvent effect, the solvent dependence of k_p should show the same pattern for both vinyl esters and methacrylates. However, this is not the case.

The polarity of the solvent radical might participate in the solvent effect. Although the polarity of the radical is difficult to be directly estimated, Hammet's constants σ_m or σ_i are considered to become a measure of the polarity of the aromatic

Table 11. Polymerization rate and elementary rate constants in the free radical polymerization of methyl methacrylate and vinyl acetate in benzene and deuterated benzene at 30 °C[17]

	Methyl methacrylate		Vinyl acetate	
	Benzene	d_6-Benzene	Benzene	d_6-Benzene
$R_p{}^a$ (mol/l · sec)	1.45×10^{-5}	1.49×10^{-5}	1.19×10^{-6}	1.13×10^{-6}
k_p (l/mol · sec)	450	456	117	113
k_t (l/mol · sec)	4.20×10^{-7}	4.38×10^{-7}	3.19×10^{-8}	3.51×10^{-8}

a [2,2'-Azobis-(isobutyronitrile)] = 0.10 mol/l, [monomer] = 2.0 mol/l

radical. However, no correlation between the Hammett parameters and k_p was found. If both stability and polarity of the radical influence the propagation rate, k_p would be linearly correlated with Hammett's σ_p and a free energy relationship would be observed. The expectations are not consistent with the experimental results, too. Accordingly, simultaneous occurrence of transfer and retardation by the transferred radical was ruled out from the elucidation of the solvent effect.

4.3 Radical Complex Between Propagating Radical and Solvent

Bamford et al.[2,3] and Burnett et al.[4,5] have explained the solvent effect on k_p for methyl methacrylate and styrene in terms of a donor-acceptor complex formation between propagating radicals and aromatic solvents. Since it was quite difficult to estimate experimentally the ability for the formation of a complex of the free radical with the solvent molecule, the delocalization stabilization for the complex formation was calculated. Generally, a free radical has an unpaired electron in a non-bonding orbital which has a low ionization potential and high electron affinity and seems to be suitable energetically for the formation of a charge-transfer complex[110]. This concept seems to explain the opposite solvent effect for vinyl esters and methacrylates in terms of the electron-accepting property of polymethacrylate radicals and the electron-donating property of poly(vinyl ester) radicals. However, since the methoxy substituent is known to increase the availability of the π-electrons in the aromatic ring, k_p for vinyl acetate should be largest in anisole. Actually, k_p in anisole is smaller than that in benzene, indicating the anomalous effect of anisole (Fig. 1). Bamford et al. and Henrici-Olivé et al. explained this anomaly on the assumption that the ether group of anisole only forms a σ-complex with the poly(methyl methacrylate) radical which is less stable than the π-complex formed between the polymer radical and benzene. However, molecular orbital calculations suggest the possibility of the formation of a π-complex of the propagating radical even in anisole, as shown in Chap. 5.

According to this concept, the electron-accepting poly(phenyl methacrylate) radical is more likely to form a π complex than the poly(methyl methacrylate) radical, because the phenyl group decreases the electron density of the propagating radical as compared to the methyl group. It is suggested that the propagating radical of methyl methacrylate is more reactive than that of phenyl methacrylate. The fact that k_p for phenyl methacrylate, which is more likely to be attacked by free radicals[96, 97], is smaller than that for methyl methacrylate supports this concept.

On the other hand, since poly(vinyl ester) radicals are reported to be electron-rich[104], they are electron donors with respect to solvent. The poly(vinyl acetate) radicals is more likely to form a complex than the poly(vinyl benzoate) radical, since the methyl group increases the electron density of the propagating radical as compared to the phenyl group. If the complexed radical is assumed to be less reactive than the free radical or inactive, the concept of the π-complex along with the experimental evidence of similar methyl affinity[47] of both monomers explains the fact that k_p for vinyl acetate is smaller than that for vinyl benzoate.

4.4 Interaction Between Monomer and Solvent

Bamford et al.[2] showed that there was no interaction between methyl methacrylate and bromobenzene, because any deviation from Raoult's law was not observed in mixtures of the two components. Raoult's law was confirmed for solutions of vinyl acetate and methyl methacrylate in benzene, ethyl acetate, acetonitrile, and acetone[108]. However, Allen et al.[109] investigated the effect of certain aromatic compounds on the ^{1}H-NMR spectra of methyl methacrylate, indicating that an interaction between monomer and solvent occurs. This result suggests that the interaction is likely to slightly affect k_p.

The NMR data of vinyl compounds are shown in Tables 12 and 13. A change in the chemical shift of the olefinic proton in the trans position relative to the ester group has been observed, whereas the chemical shift of the cis olefinic proton is scarcely influenced. The change in the chemical shift of the trans olefinic proton in the monomer increases in the following order:

Vinyl acetate = vinyl benzoate < methyl methacrylate ≈ t-butyl methacrylate < phenyl methacrylate.

As the solvent dependence of the chemical shift of methyl methacrylate is almost the same as that of t-butyl methacrylate bearing the bulky ester group, the interaction between the carbonyl or alkoxy group and the solvent molecule does not seem to be important. These results show that the trans olefinic proton interacts with the aromatic ring. This interaction might participate in the solvent effect on k_p, especially in the polymerization of methacrylates whose k_p values slightly change with solvent. In order to estimate the interaction, the NMR spectra were taken in different molar solvent ratios of benzene to cyclohexane at constant concentration of methyl methacrylate ([Monomer] = 9.0×10^{-2} M), because cyclohexane is an inert solvent. The stability constant (K) is estimated by Eq. (4.1)[110],

$$\frac{1}{\Delta} = \frac{1}{K \Delta_0 [D]} + \frac{1}{\Delta_0} \tag{4.1}$$

Table 12. Chemical shifts of olefinic protons for methacrylate monomers in various solvents [a,b 15]

Methyl methacrylate: H_b, CH_3 / $C=C$ / H_a, $COOCH_3$

t-Butyl methacrylate: H_b, CH_3 / $C=C$ / H_a, $COOC(CH_3)_3$

	δH_a (Hz)	δH_b (Hz)	$\Delta\delta^c$ (Hz)	δH_a (Hz)	δH_b (Hz)	$\Delta\delta^c$ (Hz)
Anisole	365.5	319	44.5	363	318	45
Benzene	365	314	51	364.5	312.5	52
Fluorobenzene	364.5	321	43.5	364.5	320	44.5
Chlorobenzene	363	320.5	42.5	363	319.2	43.8
Bromobenzene	363.5	321.5	42	363	320	43
Benzonitrile	365	331	34	364	328.7	35.3

[a] Tetramethylsilane was used as an internal reference. (SH_a and SH_b from tetramethylsilane)
[b] In 5% solution
[c] $\Delta\delta = \delta H_a - \delta H_b$

Table 13. Chemical shifts of olefinic protons for vinyl compounds in various solvents [a, b 14, 15]

Phenyl methacrylate: H_b, CH_3 / $C=C$ / H_a, $COOC_6H_5$

Vinyl benzoate: H_b, H / $C=C$ / H_a, $OOC-C_6H_5$

	δH_a (Hz)	δH_b (Hz)	$\Delta\delta^c$ (Hz)	δH_a (Hz)	δH_b (Hz)	$\Delta\delta^c$ (Hz)
Anisole	376.3	328	48.5	297	270	27
Benzene	377	321	56	293	262	31
Fluorobenzene	378	330	48	298	271	27
Chlorobenzene	377	330	47	298	270	28
Bromobenzene	377	331	46	298	271	27
Ethyl benzoate	380	338	42	305	280	25
Benzonitrile	382	344	38	308	284	24

[a-c] See Table 12

where Δ is the observed shift for the trans olefinic proton in the mixture relative to that in cyclohexane, Δ_0 denotes the shift for the pure interacting species relative to that in cyclohexane, and [D] the concentration of benzene (Fig. 3). In addition, $\Delta H \approx 0$ kcal/mol was found from the temperature dependence of K, being consis-

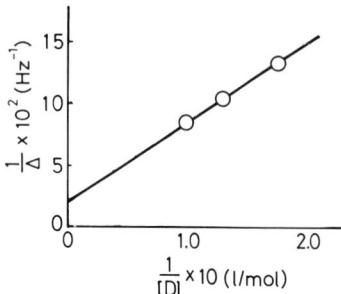

Fig. 3. Plot of $1/\Delta$ against $1/[D]$ for the π complex of methyl methacrylate with benzene in cyclohexane (Δ, D: see text). K = 0.06 l/mol

tent with the fact that deviation from Raoult's rule was not observed for methyl methacrylate in bromobenzene. This result suggests that the interaction is caused by the change of entropy through the collision complex. The energy of this interaction is too small to explain the kinetic solvent effect on the free radical polymerization in aromatic solvents. A weaker but similar solvent interaction was also observed for vinyl esters. Since the kinetic solvent effect between methacrylates and vinyl esters are opposite to the Hammett-like plot in Fig. 1, the interaction between monomer and solvent is not considered to be an important factor for the solvent effect on k_p.

4.5 Interaction between Polymer and Solvents

The solvent effect on k_p might be caused by the specific interaction between polymer and solvent (selective solvation of polymer), because the concentration of mono-

Table 14. Correlation of k_p values and the difference of solubility parameters between solvent and polymer

Solvent	MMA k_p (l/mol · sec)	δ Solvent (cal/cm^3)$^{0.5}$	δ Solvent-δ Polymer[a] (cal/cm^3)$^{0.5}$
Benzene	450	9.2	0.05
Chlorobenzene	498	9.5	0.25
Benzonitrile	614	8.4	0.85

Solvent	VAc k_p (l/mol · sec)	δ Solvent (cal/cm^3)$^{0.5}$	δ Solvent-δ Polymer[b] (cal/cm^3)$^{0.5}$
Benzene	117	9.2	0.2
Chlorobenzene	61	9.5	0.1
Benzonitrile	8	8.4	1.0

[a] Poly(methyl methacrylate) 9.25[111])
[b] Poly(vinyl acetate) 9.4[111])

mer around the propagating polymer radical would vary with the nature of solvent. When polymer interacts more strongly with monomer than with solvent, the monomer concentration around the polymer would increase and the apparent k_p value would become larger. The interaction between polymer and solvent was estimated by the difference of the solubility parameter between polymer and solvent[111]. If the interaction of solvent with polymer contributes to the solvent effect on k_p, k_p would decrease in the solvent in which the difference in the solubility parameter between polymer and solvent is smaller than that between polymer and monomer. Although the data available for the solvents used are inadequate to draw conclusions, the variation of the solubility parameter difference with solvent appears to be correlated with that of k_p in the polymerization of methyl methacrylate (Table 14). However, the solvent effect on k_p for vinyl acetate cannot be explained by the difference in the solubility parameter between solvent and poly(vinyl acetate). Therefore, it is clear that the interaction between solvent and polymer is not considered to be an important factor for the solvent effect on k_p in the polymerization of vinyl acetate.

5 Quantum Chemical Considerations on the Radical Complex

Mulliken's[112, 113] theory implies that a charge-transfer interaction takes place when the overlap of the highest occupied molecular orbital (HOMO) of the donor with the lowest unoccupied molecular orbital (LUMO) of the acceptor is at a maximum. Fukui et al.[114, 115] applied the delocalization method to the Ag^+/arene complex. The electron stabilization energy, which is due to the delocalization from the occupied orbital of the arene to the lowest unoccupied orbital of the silver cation, was calculated by the purturbation theory. The Ag^+ ion is located above the C=C bond such that the sum of the corresponding coefficients of the HOMO has a maximum. This location is in conformity with X-ray studies. A clear parallelism was found between the delocalization energy and the stability constants of Ag^+/benzene derivatives[114]. Similarly, the quinhidrine-type complex[116] is explained by the symmetry relation of HOMO of hydroquinone and LUMO of quinone which is most favorable for the HOMO-LUMO interactions. The concept of HOMO-LUMO, HOMO-SOMO (single occupied molecular orbital) or LUMO-SOMO control can be more generally used for the prediction of stable shape of simple molecules or salts[117].

We applied this calculation to the estimation of the stability of the radical complex. Since SOMO in the free radical can play, according to the orbital energy relationship and the orbital overlapping situation, the role of HOMO, LUMO or both, the mutual charge transfer from the free radical to LUMO of solvent or from HOMO of solvent to SOMO of the free radical is of particular significance. Since the poly(vinyl ester) radical is electron-rich[104], the location of the propagating radical on the aromatic ring was assumed to be determined by SOMO-LUMO interactions. When LUMO of solvent has no adjacent atomic orbital of the same sign in the aromatic ring, the location has been determined by SOMO-second LUMO interactions. On the other hand, since polymethacrylate radicals are considered to be electron-poor, the location has been determined by HOMO-SOMO interactions, Delocalization energies for three kinds of complex formation have been estimated as delocalization stabilizations by the purturbation method using Eqs. (5.1) – (5.3)[114, 115].

Three kinds of models for the calculations are shown in Fig. 4.

The delocalization energy is expressed by the relationship

$$-\Delta E = \frac{(C_r^{LU} + C_s^{LU})^2 (d_N^{SO})^2}{\epsilon_N - \epsilon_{LU}} \gamma^2 \beta^2 + \frac{(C_r^{HO} + C_s^{HO})^2 (d_N^{SO})^2}{\epsilon_{HO} - \epsilon_N} \gamma^2 \beta^2 \quad (5.1)$$

where ϵ_{HO} and ϵ_{LU} are the energies of HOMO and LUMO of solvent, ϵ_N is the energy of the non-bonding molecular orbital of the propagating radical, and γ the resonance integral between the radical and the rth and sth atoms of the aromatic molecule. Notations C_r^{HO} and C_s^{HO}, and C_r^{LU} and C_s^{LU} are the coefficients of the rth and sth C-atom orbitals in HOMO and LUMO of solvent in which the rth and sth carbon atoms are adjacent to each other. Notation d_N^{SO} is the coefficient of the C-atom orbital of the propagating radical in the non-bonding molecular orbital. The results of the estimation for phenyl methacrylate according to Eq. (5.1) are shown in Table 15, indicating that models 2 and 3 do not explain the variation of k_p with solvent. Therefore, model 1 was assumed for the formation of the complex. In addition, three cases (weak, medium, and strong interactions) are considered in model 1. (Case 1) When the interaction between donor and acceptor is very weak, only LUMO-SOMO and HOMO-SOMO interactions are taken into account and the delocalization energy is calculated by Eq. (5.1). (Case 2) For medium interactions, both delocalization of electrons in the occupied molecular orbital of solvent with respect to SOMO and that of electrons in SOMO with respect to the unoccupied molecular orbitals of solvent are considered and the delocalization energy is calculated by Eq. (5.2).

$$-\Delta E = \sum_j^{unocc} \frac{(C_r^j + C_s^j)^2 (d_n^{so})^2}{E_N - E_j} \gamma^2 \beta^2 + \sum_i^{occ} \frac{(C_r^i + C_s^i) (d_n^{so})^2}{E_i - E_N} \gamma^2 \beta^2 \quad (5.2)$$

Table 15. Delocalization stabilization in three models for the complex formation of poly(phenylmethacrylate) radical

Solvent	model 1 ($-\gamma^2\beta$)	model 2 ($-\gamma^2\beta$)	model 3 ($-\gamma^2\beta$)	k_p (l/mol · sec)
Anisole[a]	0.526	0.485	0.849	230
Benzene	0.827	0.427	0.219	176
Fluorobenzene[b]	0.741	0.213	0.915	180
Chlorobenzene[c]	0.655	0.186	0.895	223
Bromobenzene[d]	0.628	0.203	0.862	235
Benzonitrile[e]	0.359	0.529	0.276	275

[a] Anisole $\alpha_{OCH_3} = \alpha + 0.5\beta$, $\alpha_{adj.C} = \alpha$, $\beta_{C-OCH_3} = 0.6\beta$[118, 119)
[b] Fluorobenzene $\alpha_F = \alpha + 2.1\beta$, $\alpha_{adj.C} = \alpha + 0.2\beta$, $\beta_{C-F} = 1.25\beta$[118)
[c] Chlorobenzene $\alpha_{Cl} = \alpha + 1.8\beta$, $\alpha_{adj.C} = \alpha + 0.18\beta$, $\beta_{C-Cl} = 0.8\beta$[118)
[d] Bromobenzene $\alpha_{Br} = \alpha + 1.4\beta$, $\alpha_{adj.C} = \alpha + 0.14\beta$, $\beta_{C-Br} = 0.7\beta$[118)
[e] Benzonitrile $\alpha_N = \alpha + 0.8\beta$, $\alpha_{adj.C} = \alpha + 0.08\beta$, $\beta_{C-N} = \beta$[119)

Influence of Solvent

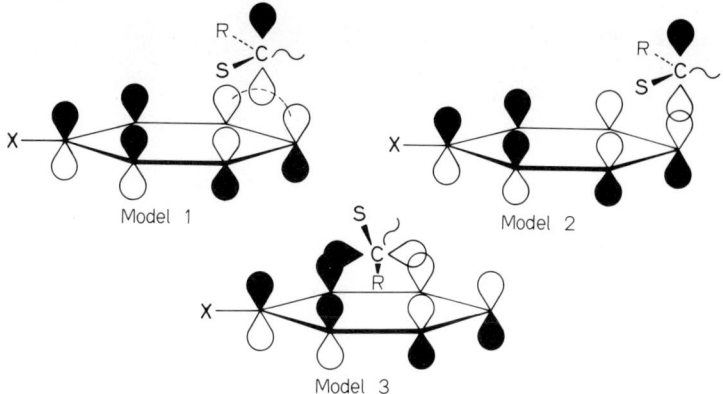

Fig. 4. Assumed models for the calculation of delocalization stabilization for the formation of complexes (see text)

(Case 3) In the case of strong interactions not only unpaired electrons but also paired electrons in the free radical possibly participate in the molecular interaction with solvent. Accordingly, the delocalization energy must be calculated taking into account the delocalization from occupied electrons to unoccupied orbitals of interesting molecules (Eq. 5.3):

$$-\Delta E = 2 \left(\sum_{i}^{occ} \sum_{j}^{unocc} - \sum_{i}^{occ-N} \sum_{j}^{unocc} \right) \frac{(d_n^i)^2 (C_r^j + C_s^j)^2}{E_i - E_j} (\gamma \beta)^2$$

$$+ \left(\sum_{i}^{occ} - \sum_{j}^{unocc} \right) \frac{(d_n^N)^2 (C_r^j + C_s^j)^2}{E_N - E_j} (\gamma \beta)^2 \quad (5.3)$$

The variation of the delocalization energy with solvent is in best conformity with that of k_p in case 1 for methacrylates, in case 2 for vinyl benzoate and in case 3 for vinyl acetate. The results of the calculations are shown in Table 16 and Fig. 5, indicating that there is a good correlation between k_p and the delocalization energy.

Table 16. Delocalization stabilization for complex formation of propagating radicals with aromatic solvents

Solvent	Phenyl methacrylate $(-\gamma^2\beta)$	Vinyl benzoate $(-\gamma^2\beta)$	Vinyl acetate $(-\gamma^2\beta)$
Anisole[a]	0.526	1.219	1.466
Benzene[a]	0.827	1.180	1.392
Fluorobenzene[a]	0.741	1.177	1.419
Chlorobenzene[a]	0.655	1.185	1.417
Bromobenzene[a]	0.628	–	–
Ethyl benzoate[b]	–	1.290	1.574
Benzonitrile[a]	0.359	1.366	1.723

[a] Table 15
[b] Ethyl benzoate, $\alpha_O = \alpha + 1.0 \beta$, $\alpha_{COO} = \alpha + 0.2 \beta$, $\alpha_{adj.C} = \alpha$, $\beta_{C=O} = \beta$, $\beta_{C-O} = 0.6 \beta$[119])

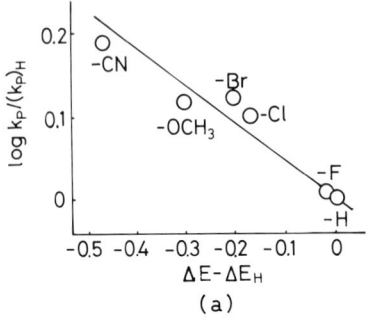

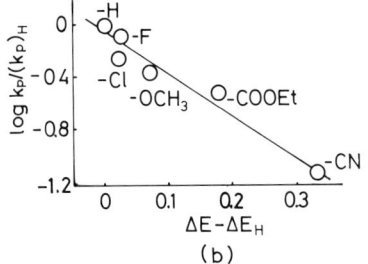

Fig. 5a, b. Correlation of k_p and delocalization stabilization in various solvents. $(k_p)_H : k_p$ in benzene. (a) Phenyl methacrylate, (b) vinyl acetate

Molecular orbital calculations for the complex formation between propagating radical and solvent indicate that the effect of anisole is not anomalous, and that the likelihood of the complex formation is in the following order:
vinyl acetate > vinyl benzoate > phenyl methacrylate ≈ methyl methacrylate.

This order is consistent with other experimental results. The methyl group at the α-position of the propagating end in the polymethacrylate radical is considered to sterically hinder the complex formation, and hence, even if the complex is formed, the complex is possibly very weak, the ability of the complex formation of the polymethacrylate radical being smaller than that of the poly(vinyl ester) radical. The difference between the poly(vinyl acetate) radical and the poly(vinyl benzoate) radical might be caused by the polarity and the bulkiness of the ester group, because the phenyl group is a weaker electron donor and bulkier than the methyl group. The results of the molecular orbital method indirectly support the concept that some of the propagating radicals might be stabilized by the complex formation with aromatic solvents.

6 Mechanistic View on Free Radical Polymerization of Vinyl Esters in Aromatic Solvents

The influence of aromatic solvents on k_p for all the monomers used is most reasonably explained in terms of the donor-acceptor complex between the propagating radical and the aromatic solvent.

Thus, the behavior of the propagating radical can be represented by Eqs. (6.1) – (6.3):

$$P_n\cdot + S \underset{}{\overset{K_s}{\rightleftarrows}} P_n\cdot S \qquad (6.1)$$

$$P_n\cdot + M \underset{}{\overset{K_M}{\rightleftarrows}} P_n\cdot M \qquad (6.2)$$

$$P_n\cdot + M \rightleftarrows P_{n+1}\cdot \qquad (6.3)$$

The propagating radical forms a complex with the aromatic ring of the solvent and with the monomer according to Eqs. (6.1) and (6.2), respectively. The complex formation of the propagating radical through the C=C bond of the monomer must also be considered and this complexation is assumed to inevitably lead to propagation. Complexed radicals $P_s \cdot S$ and $P \cdot M$ are considered to be in a dormant state. The rate constant is given by

$$k_p = \frac{k_{po}}{1 + K_M[M] + K_s[S]} \qquad (6.4)$$

Since the polymerization system lacks an aromatic ring in the polymerization of vinyl acetate in ethyl acetate, K_M and K_s can reasonably be regarded as zero, and k_p is considered to be k_{po}. Accordingly, k_p for vinyl acetate is given by Eq. (6.5) and the stability constant in aromatic solvents is estimated from Eq. 6.5 using k_p in the corresponding solvent and k_{po}. The value K_s in aromatic solvents are shown in Table 17, being clearly correlated with the delocalization

Table 17. Correlation between delocalization stabilization and stability constant for the formation of complexes[116]

Solvents	ΔE $(-\gamma^2\beta)$	K_s (l/mol)
Anisole	1.466	1.63
Benzene	1.392	0.48
Fluorobenzene	1.419	0.64
Chlorobenzene	1.417	1.17
Ethyl benzoate	1.574	2.87
Benzonitrile	1.723	9.82

Table 18. k_p values in mixed solvents of ethyl acetate (EtAc) and ethyl benzoate (EtBz)[121]

Solvent EtAc : EtBz (Molar ratio)	Observed[a] k_p (l/mol · sec)	Calculated[b] k_p (l/mol · sec)
8.22 : 0	637	–
5.48 : 1.88	104	100
4.11 : 2.82	78	70
2.74 : 3.76	48	54
0 : 5.64	37	–

[a] Polym. temp. 30 °C; [Monomer] = 2.0 mol/l
[b] K_s = 2.87 l/mol was used for calculations

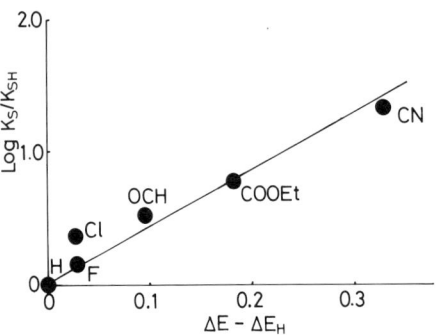

Fig. 6. Correlation of the stability constant for the poly(vinyl acetate) radical with the delocalization stabilization in various solvents

$$k_p = \frac{k_{po}}{1 + K_s[S]} \tag{6.5}$$

stabilizations for the complex (Fig. 6).

In order to check whether this mechanism is reasonable or not, k_p in mixed solvents of ethyl acetate and ethyl benzoate was estimated and compared with k_p calculated from Eq. (6.5) using k_{po} and K_s in ethyl benzoate. The calculated values are in fair agreement with the observed values (Table 18).

When k_p in mixed solvents is calculated from the k_p values in both solvents by their proportional allotment, the calculated values are not in accordance with the observed values (Table 18). The agreement in Eq. (6.5) supports our view that the complexes radical is in a dormant state in propagation. Recently, Yamamoto et al.[90] reported that the stereoregularity of poly(vinyl acetate) did not vary with these solvents. This result is consistent with the assumption that the complexed radical does not participate in the propagation process, because stereoregularity is considered to be affected by solvent if two kinds of propagating species, the free and complexed one, can add to the monomer.

Similarly, the variation of k_p with solvent in methacrylate polymerization can be explained on the assumption that the complexed radical is either inactive or less reactive. Since methyl methacrylate has no aromatic ring in itself, it seems to be possible to estimate the stability constants for the complex formation of the poly(methyl methacrylate) radical end with aromatic solvents. However, since the variation of k_p in methyl methacrylate polymerization with solvent is too small, the determination of K_s with significant figures is impossible. Accordingly, it is difficult to estimate the unpurturbed k_{po} value for methyl methacrylate and thus difficult to estimate the stability constant of the complex in aromatic solvents.

7 Solvent Effect on Free Radical Copolymerization

The polarity of the double bond resulting from the electronegative or electropositive or electropositive character of the substituent is one of the most important factors in the determination of the copolymer composition[22, 123, 126]. In spite of this polar effect, it has been uniformly observed that monomer reactivity ratios are insensitive to the nature of the medium[24–28]. Overberger et al.[127] indicated that the copolymerization reactivity of methacrylic acid both with 2-diethylaminoethyl methacrylate and with acrylonitrile markedly depends upon pH of the polymerization system. This is interpreted in terms of the ionization of methacrylic acid at high pH. This kind of solvent effect is evident in systems including ionogenic monomers such as acrylic acid and sodium styrenesulfonate[128, 131].

Keber[129] correlated the changes in the reactivity ratio of styrene to acrylic acid with associations of the solvent with acrylic acid through hydrogen bonding. The solvent effect on the monomer reactivity ratio was also observed in the free radical copolymerization of non-ionizable monomers such as acrylamide, N-monosubstituted acrylamides, methacrylamide, acrylonitrile, and methacrylonitrile[132, 141]. Saini et al.[132-135] reported that the monomer reactivity ratio for the copolymerization of acrylamide with styrene is strongly affected by the solvent and considered to be affected by the degree of ketoenol equilibrium according to Eq. (7.1):

$$CH_2=CH-\underset{O}{\underset{\|}{C}}-NH_2 \longleftrightarrow CH_2=CH-\underset{O^-}{\underset{|}{C}}=\overset{+}{N}H_2 \rightleftharpoons CH_2=CH-\underset{OH}{\underset{|}{C}}=NH \qquad (7.1)$$

This view was supported by the observation that the reactivity of N-methylacrylamide is similarly influenced by the medium, while that of N,N-dimethylacrylamide, which cannot enolize according to scheme (7.1), is not. On the other hand, Chatterjee et al.[139] reported that the change in the reactivity of acrylamide is attributable to hydrogen bonding of the N–H bond. They carried out the copolymerization of acrylamide and N-vinylpyrrolidone in binary aqueous solutions containing 0, 30, 67, and 80% glycol by weight. The reactivity of acrylamide relative to that of N-vinylpyrrolidone increases at higher glycerol concentration. They considered that the presence of glycerol enhances the hydrogen bonding of the acrylamide monomer with the polymer and in particular with the polymer radical.

In the end of 1960s, Nikolaev et al.[29] and Ito et al.[30] independently demonstrated an appreciable effect of the reaction medium on the reactivity ratios in the copolymerization of methyl methacrylate and styrene (Table 19). Ito et al. found that the relative reactivity of methyl methacrylate toward the polystyryl radical is correlated with the transition energies E_T for the longest wavelength absorption band for pyridinum N-phenolbetaine in solvents. They suggested that the polarized structure of methyl methacrylate monomer becomes important in the transition state. Bonta et al.[32] also demonstrated that there is an appreciable solvent effect on the reactivity ratio in the styrene-methyl methacrylate copolymerization in non-

Table 19. Reactivity ratios in the copolymerization of styrene (M_1) and methyl methacrylate (M_2) in different solvents[30]

Solvent	r_1	r_2
Benzene	0.57 ± 0.032	0.46 ± 0.032
Benzonitrile	0.48 ± 0.045	0.49 ± 0.045
Benzyl alcohol	0.44 ± 0.054	0.39 ± 0.054
Phenol	0.35 ± 0.024	0.35 ± 0.024

aromatic solvents. They indicated that both reactivity ratios decrease with increasing solvent dielectric constant.

The growing chain ending with a styrene unit is mainly negatively polarized by the electron-donating phenyl group whereas the electron-accepting ester group gives positively polarized chain ends. They considered that polar solvents would enhance the role of the polarized form, resulting in a marked increase of the alternating tendency. Similarly, the solvent effect on the reactivity ratio has been observed in styrene-acrylamide[140], styrene-methacrylonitrile[33], methyl methacrylate-acrylonitrile[131], styrene-acrylonitrile[141], and styrene-pyridazinone copolymerizations[34, 35].

Chapiro[142] has drawn attention to the fact that physical aggregation phenomena are often neglected in the interpretation of copolymerization data. He showed that copolymerizations involving monomers which can form aggregates exhibit very pronounced solvent effects. The importance of such effects was pointed out by Takemoto et al.[143, 144], the copolymerization of a vinyl monomer having a nucleic acid base.

Let us consider the copolymer composition in the light of the complex formation, taking into account eight propagation reactions:

$$
\begin{aligned}
\sim M_1\cdot + M_1 &\xrightarrow{k_{11}} \sim M_1M_1\cdot \text{ or } \sim M_1M_1\cdot S \\
\sim M_1\cdot S + M_1 &\xrightarrow{k'_{11}} \sim M_1M_1\cdot \text{ or } \sim M_1M_1\cdot S \\
\sim M_1\cdot + M_2 &\xrightarrow{k_{12}} \sim M_1M_2\cdot \text{ or } \sim M_1M_2\cdot S \\
\sim M_1\cdot S + M_2 &\xrightarrow{k'_{12}} \sim M_1M_2\cdot \text{ or } \sim M_1M_2\cdot S \\
\sim M_2\cdot + M_1 &\xrightarrow{k_{21}} \sim M_2M_1\cdot \text{ or } \sim M_2M_1\cdot S \\
\sim M_2\cdot S + M_1 &\xrightarrow{k'_{21}} \sim M_2M_1\cdot \text{ or } \sim M_2M_1\cdot S \\
\sim M_2\cdot + M_2 &\xrightarrow{k_{22}} \sim M_2M_2\cdot \text{ or } \sim M_2M_2\cdot S \\
\sim M_2\cdot S + M_2 &\xrightarrow{k'_{22}} \sim M_2M_2\cdot \text{ or } \sim M_2M_2\cdot S
\end{aligned} \quad (7.2)
$$

Two equations must be considered in the formation of the complex

$$
\begin{aligned}
\sim M_1\cdot + S &\overset{K_1}{\rightleftarrows} \sim M_1\cdot S \\
\sim M_2\cdot + S &\overset{K_2}{\rightleftarrows} \sim M_2\cdot S
\end{aligned} \quad (7.3)
$$

The concentration of the uncomplexed radical is given by

$$
[M_1\cdot] = \frac{1}{1 + K_1[S]} [M_1\cdot]_0 = \alpha [M_1\cdot]_0
$$

$$
[M_2\cdot] = \frac{1}{1 + K_2[S]} [M_2\cdot]_0 = \beta [M_2\cdot]_0
$$
(7.4)

where α and β are fractions of the uncomplexed radical, and $[M_1\cdot]_0$ and $[M_2\cdot]_0$ are total concentrations of the propagating radical of M_1 and M_2, respectively. The molar ratio of monomers in the copolymer is

Influence of Solvent

$$\frac{d[M_1]}{d[M_2]} = \frac{k_{11}[M_1\cdot][M_1] + k'_{11}[M_1\cdot S] + k_{21}[M_2\cdot][M_1] + k'_{21}[M_2\cdot S][M_1]}{k_{12}[M_1\cdot][M_2] + k'_{12}[M_2\cdot S] + k_{22}[M_2\cdot][M_2] + k'_{22}[M_2\cdot S][M_2]} \quad (7.5)$$

Assuming a stationary state, the radical concentration is

$$-\frac{d[M_1\cdot]}{dt} = -\frac{d[M_1\cdot S]}{dt} = -\frac{d[M_2\cdot]}{dt} = -\frac{d[M_2\cdot S]}{dt} = 0 \quad (7.6)$$

Then,

$$(k_{12}[M_1\cdot] + k'_{12}[M_1\cdot S])[M_2] = (k_{21}[M_2\cdot] + k'_{21}[M_2\cdot S])[M_1] \quad (7.7)$$

Eq. (7.7) can be rearranged by using Eq. (7.4) and combined with Eq. (7.5) to yield

$$\frac{d[M_1]}{d[M_2]} = \frac{\left(\dfrac{\alpha k_{11} + k'_{11}(1-\alpha)}{\alpha k_{12} + k'_{12}(1-\alpha)}[M_1] + [M_2]\right)}{\left([M_1] + \dfrac{k_{22}\beta + k'_{22}(1-\beta)}{k_{21}\beta + k'_{21}(1-\beta)}[M_2]\right)} \frac{[M_1]}{[M_2]} \quad (7.8)$$

If the complexed radical is inactive ($k'_{11} = k'_{12} = k'_{22} = k'_{21} = 0$), Eq. (7.8) reduces to the ordinary Mayo-Lewis equation and no solvent effect on the reactivity ratio will be observed. Busfield et al.[108] studied the solvent effect on the free radical copolymerization of vinyl acetate and methyl methacrylate. The methyl methacrylate content is unaffected by benzene and ethyl acetate. This result seems to be consistent with our assumption that the complexed radical is inactive in propagation. However, the solvent effect might not be observed in the case in which the reactivity of the complexed radical is proportional to that of the uncomplexed radical, because also in this case Eq. (7.8) reduces to the Mayo-Lewis form. It is difficult, therefore, to expect from the copolymerization experiment some evidence to support the concept of the complex formation.

8 Concluding Remarks

In order to illustrate the influence of aromatic solvents on the radical polymerization rate, the elementary rate constants of the polymerization of vinyl esters and methacrylates were determined in various aromatic solvents. The solvent effect on the polymerization rate for vinyl esters is mainly caused by propagation processes only, whereas that for methacrylates is due to both the termination and to propagation processes. In the Hammett plot, the variations of propagation rate constants of vinyl ester with the substituents of the aromatic solvents are inverse to those of methacrylate polymerization. This effect was most reasonably explained in terms of transient donor-acceptor complex between the propagating radical and solvent. The concept suggests that varying amounts of the propagating radicals are temporarily stabilized as complexes. At present, however, this concept represents a proposal in the molecular interpretation of the solvent effect. It will be necessary to confirm the formation of the radical complex either by utilizing a new concept or by the application of a more ingenious technique. Application of ESR, flash photolysis and a combination of the two techniques to the radical polymerization system will offer a more detailed knowledge about the nature of the growing species. If the radical complex is confirmed to be a reaction intermediate, it will afford new information on the reaction control of free radical polymerization. In this article, the influence of metal salts on the polymerization rate was not mentioned. Recently, the possibility of free radical living polymerizations of vinyl compounds in homogeneous systems in the presence of metal salts[145, 146] has been reported. This might be an extreme case of the formation of a radical complex.

Acknowledgement. The author wishes to express his sincere appreciation to Professor Shun-ichi Nozakura for his frequent discussions and encouragement. He is also grateful to Professor Der Jang Liaw (Taiwan Institute Technology) and Mr. Jiro Satoh (Asahi Chemical Co.) whose skilled experimental assistance has contributed to write this manuscript.

9 References

1. Flory, P. J.: Principles of Polymer Chemistry, pp. 106–177. Ithaca, NY: Cornell University Press 1953
2. Bamford, C. H., Brumby, S.: Makromol. Chem. *105,* 222 (1967)
3. Bamford, C. H.: Radical Intermediates in Radical Polymerization Processes. In: Ledwith, A. and North, A. (eds.): Molecular Behavior and the Development of Polymer Materials, pp. 51–87, London: Chapman and Hall 1975
4. Burnett, G. M., Cameron, G. G., Zafar, M. M.: Europ. Polym. J. *6,* 823 (1970)
5. Burnett, G. M., Cameron, G. G., Joiner, S. N.: J. Chem. Soc., Faraday Trans. I, *69,* 322 (1973)
6. Burnett, G. M., Cameron, G. G., Parker, B. M.: Europ. Polym. J. *5,* 231 (1969)
7. Yamamoto, T. et al.: Nippon Kagaku Kaishi, *1978,* 1516
8. Yamamoto, T., Yamamoto, T.: Rep. Himeji Inst. Tech. *25,* A, 126 (1972)
9. Zafar, M. M.: Makromol. Chem. *157,* 219 (1972)
10. Bamford, C. H., Brumby, S.: Chem. & Ind. *1969,* 1020
11. Mayer, G., Schulz, G. V.: Makromol. Chem. *173,* 101 (1973)
12. Fischer, J. P., Mucke, G., Schulz, G. V.: Ber. Bunsenges. Phys. Chem. *73,* 154 (1969)
13. Fischer, J. P., Mucke, G., Schulz, G. V.: Ber. Bunsenges. Phys. Chem. *74,* 1077 (1970)
14. Kamachi, M., Liaw, D. J., Nozakura, S.: Polym. Prepr. *20,* 567 (1979)
15. Kamachi, M., Liaw, D. J., Nozakura, S.: Polym. J. *9,* 307 (1977)
16. Kamachi, M., Satoh, J., Nozakura, S.: J. Polym. Sci., Polym. Chem. Ed. *16,* 1789 (1978)
17. Kamachi, M., Liaw, D. J., Nozakura, S.: Polym. J. *11,* 921 (1979)
18. Norrish, G. W., Brookman, E. F.: Proc. Royal Soc., London(A), *171,* 147 (1939)
19. Schulz, G. V., Dinglinger, A., Huseman, E.: Z. Physik. Chem.(B) *43,* 385 (1939)
20. Schulz, G. V., Blascheke, F.: Z. Elektrochem. *47,* 749 (1941)
21. Tromsdorff, E., Kohle, H., Lagally, P.: Makromol. Chem. *1,* 106 (1947)
22. Bengough, W. I., Mellville, H. W.: Proc. Royal Soc., London(A), *230,* 429 (1955)
23. Hayden, P., Melville, H. W.: J. Polym. Sci. *43,* 201 (1960)
24. Alfrey, T., Price, C. C.: J. Polym. Sci. *2,* 101 (1947)
25. Price, C. C., Walsh, J. G.: J. Polym. Sci. *6,* 239 (1951)
26. Walling, C. and Mayo, F. R.: J. Polym. Sci. *3,* 895 (1948)
27. Lewis, F. et al.: J. Am. Chem. Soc. *70,* 1519 (1948)
28. Price, C. C.: J. Polym. Sci. *3,* 772 (1948)
29. Nikolaev, A. F., Galperin, V. M.: Vysokomol. Soedin. *9A,* 2469 (1967)
30. Itoh, T., Otsu, T.: J. Macromol. Sci., J. Chem. *A3,* 197 (1969)
31. Zafar, M. M., Mahmud, R., Syed, A. M.: Makromol. Chem. *175,* 1531 (1974)
32. Bontà, G., Gallo, B. M., Russo, S.: Polymer *16,* 429 (1975)
33. Cameron, G. G., Esselmont, G. F.: Polymer *13,* 435 (1972)
34. Matsubara, Y. et al.: J. Polym. Sci., Polym. Chem. Ed. *13,* 913 (1975)

35. Matsubara, Y., Yoshihara, M., Maeshima, T.: J. Polym. Sci., Polym. Chem. Ed. *14*, 896 (1976)
36. Burnett, G. M., Melville, H. W.: Discuss. Faraday Soc. *2*, 322 (1947)
37. Conix, A., Smets, G.: J. Polym. Sci. *10*, 525 (1953)
38. Burnett, G. M., Loan, L. D.: Trans. Faraday Soc. *51*, 214, 219, 226 (1955)
39. Jenkins, A. D.: Trans. Faraday Soc. *54*, 1885, 1895 (1958)
40. Jenkins, A. D.: J. Polym. Sci. *29*, 245 (1958)
41. Stockmayer, W. H., Peebles, Jr., L. H.: J. Am. Chem. Soc. *75*, 2278 (1953)
42. Peeble, Jr., L. H., Clarke, J. T., Stockmayer, W. H.: J. Am. Chem. Soc. *82*, 4780 (1960)
43. Breitenbach, J. W.: Monatsh. Chem. *92*, 1100 (1961)
44. Mortimer, G. A., Arnold, L. C.: J. Am. Chem. Soc. *84*, 4986 (1962)
45. Henrici-Olivé, G., Olivé, S.: Makromol. Chem. *51*, 236 (1962)
46. Haas, H. C., Husek, H.: J. Polym. Sci. *A2*, 2297 (1964)
47. Santee, G. F. et al.: Makromol. Chem. *73*, 177 (1964)
48. Mayo, F.: J. Am. Chem. Soc. *75*, 6133 (1953)
49. Russell, G. A.: J. Am. Chem. Soc. *80*, 4987 (1958)
50. Russell, G. A.: Tetrahedron *8*, 101 (1960)
51. Russell, G. A., Ito, A., Hendry, D. G.: J. Am. Chem. Soc. *85*, 2976, 4997 (1963)
52. Strong, R. L., Perano, J.: J. Am. Chem. Soc. *83*, 2843 (1961)
53. Strong, R. L., Rand, S. J., Britl, A. J.: J. Am. Chem. Soc. *82*, 5053 (1960)
54. Strong, R. L., J. Phys. Chem. *66*, 2423 (1962)
55. Martin, M. M., Gleicher, G. J.: J. Am. Chem. Soc. *86*, 238 (1964)
56. Russell, G. A.: J. Org. Chem. *24*, 300 (1959)
57. Vrancken, A., Smets, G.: Makromol. Chem. *30*, 197 (1959)
58. Morrison, E. D., Gleason, E. H., Stannett, V.: J. Polym. Sci. *36*, 267 (1959)
59. Smets, G., Hertoghe, A.: J. Polym. Sci. *17*, 189 (1956)
60. Burnett, G. M., Wright, W. W.: Trans. Faraday Soc. *49*, 1108 (1953)
61. Banerjee, S., Muthana, M.: J. Polym. Sci. *37*, 467 (1953)
62. Litt, M., Stannett, V.: Makromol. Chem. *37*, 19 (1960)
63. Anderson, D. B., Burnett, G. M., Gowan, A. C.: J. Polym. Sci. *A1*, 1465 (1963)
64. Burnett, G. M., Dailey, W. S., Pearson, J. M.: Trans. Faraday Soc. *61*, 1216 (1965)
65. Henrici-Olivé, G., Olivé, S.: Makromol. Chem. *58*, 188 (1962)
66. Burnett, G. M., Dailey, W. S., Pearson, J. M.: Europ. Polym. J. *5*, 231 (1969)
67. Henrici-Olivé, G., Olivé, S.: Makromol. Chem. *68*, 219 (1963)
68. Henrici-Olivé, G., Olivé, S.: Z. Phys. Chem. *47*, 286 (1965)
69. Henrici-Olivé, G., Olivé, S.: Z. Phys. Chem. *48*, 35 (1966)
70. Henrici-Olivé, G., Olivé, S.: Z. Phys. Chem. *48*, 51 (1966)
71. Henrici-Olivé, G., Olivé, S.: Makromol. Chem. *96*, 221 (1966)
72. Bengough, W. I., Henderson, N. K.: Chem. S. & Ind. *1969*, 657
73. Zafar, M. M., Mahmud, R.: Makromol. Chem. *175*, 2627 (1974)
74. Yokota, K., Kondo, A.: Makromol. Chem. *171*, 113 (1973)
75. Kalashnikova, L. A. et al.: Russ. J. Phys. Chem. *43*, 31 (1969)
76. Kalashnikova, L. A., Neiman, M. B., Buchachenko, A. L.: Russ. Phys. Chem. *42*, 598 (1968)
77. Buchachenko, A. L. et al. Kinetika i kataliz. *6*, 601 (1965)
78. Burnett, G. M., Cameron, G. G., Cameron, J.: Trans. Faraday Soc., *69*, 864 (1973)
79. Fox, T. G., Schnecko, H. W.: Polymer *3*, 575 (1962)
80. Watanabe, H., Sono, Y.: Kogyo Kagaku Zasshi *65*, 273 (1962)
81. Schröder, G.: Makromol. Chem. *97*, 232 (1966)
82. Elias, H., Geoldi, P., Kamat, V. S.: Makromol. Chem. *117*, 269 (1968)
83. Elias, H.: Makromol. Chem. *137*, 277 (1970)
84. Elias, H., Geoldi, P.: Makromol. Chem. *144*, 85 (1971)
85. Elias, H., Riva, M., Geoldi, P.: Makromol. Chem. *145*, 163 (1971)
86. Geoldi, P., Elias, H.: Makromol. Chem. *153*, 81 (1972)
87. Yamada, A.: Kogyo Kagaku Zasshi *73*, 2265 (1970)
88. Yamada, A. et al.: Kogyo Kagaku Zasshi *73*, 2352 (1970)

89. Nozakura, S. et al.: J. Polym. Sci., Polym. Chem. *11*, 279 (1973)
90. Yamamoto, T. et al.: Kobunshi Ronbunshu *36*, 559 (1979)
91. Sumi, M., Imoto, M.: Makromol. Chem. *50*, 161 (1961)
92. Imoto, M., Takemoto, K., Nakai, K.: Makromol. Chem. *48*, 80 (1961)
93. Rosen, J., Burleigh, P. H., Gillespie, J. H.: J. Polym. Sci. *54*, 31 (1961)
94. Burnett, G. M., Ross, F. L., Hay, J. N.: J. Polym. Sci. A1, *5*, 1467 (1967)
95. Arimoto, F. S.: J. Polym. Sci. A1, *4*, 275 (1966)
96. Kamachi, M., Liaw, D. J., Nozakura, S.: Polym. J. *10*, 641 (1978)
97. Otsu, T., Yamada, B., Sugiyama, S.: Kobunshi Ronbunshu *35*, 705 (1978)
98. Benson, S. W., North, A. M.: J. Am. Chem. Soc. *81*, 1339 (1959)
99. North, A. M., Reed, G. A.: Trans. Faraday Soc. *57*, 859 (1961)
100. North, A. M., Reed, G. A.: J. Polym. Sci. *A1*, 1311 (1963)
101. Benson, S. W., North, A. M.: J. Am. Chem. Soc. *84*, 935 (1962)
102. Yokota, K., Itoh, M.: J. Polym. Sci. B *6*, 825 (1968)
103. Yamamoto, T. et al.: Kobunshi Ronbunshu *36*, 625 (1979)
104. Kinoshita, M., Irie, T., Imoto, M.: Makromol. Chem. *110*, 47 (1973)
105. Tsuda, K., Kobayashi, S., Otsu, T.: J. Polym. Sci. A1, *6*, 41 (1968)
106. Otsu, T.: Structure and Reactivity of Vinyl Monomers in Radical Polymerization. In: Imoto, M. and Onogi, S. (eds.): Progress in Polymer Science. Vol. 1, pp. 1–63, Tokyo: Kodansha 1971
107. Clarke, J. T., Howard, R. O., Stockmayer, W. H.: Makromol. Chem. *44/46*, 427 (1961)
108. Busfield, W., Low, R. B.: Europ. Polym. J. *11*, 309 (1975)
109. Allen, P. E. M., Bateup, B. O.: Europ. Polym. J. *9*, 1283 (1973)
110. Forster, R.: Organic Charge Transfer Complexes, p. 142, London – New York: Academic Press 1969
111. Burrell, H.: Solubility Parameter Values. In: Brandrup, J. and Immergut, E. H. (eds.): Polymer Handbook, Vol. IV, pp. 337–359, New York: Interscience 1975
112. Mulliken, R. S.: J. Am. Chem. Soc. *74*, 811 (1952)
113. Mulliken, R. S.: Rec. Trav. Chim. Pays Bas *75*, 845 (1956)
114. Fukui, K. et al.: Bull. Chem. Soc., Japan, *34*, 1076 (1961)
115. Fukui, K.: Theory of Orientation and Stereoselection, pp. 19–21, Berlin, Heidelberg, New York: Springer-Verlag 1975
116. Tsubomura, H.: Bull. Chem. Soc., Japan, *26*, 304 (1953)
117. Walsh, A. D.: J. Chem. Soc. *1953*, 2260, 2266, 2288
118. O'driscoll, K. F., Yonezawa, T.: Rev. Macromol. Chem. *1*, 1 (1966)
119. Kawabata, N., Tsuruta, T., Furukawa, J.: Makromol. Chem. *51*, 80 (1961)
120. Kamachi, M. et al.: Macromolecules *10*, 501 (1977)
121. Kamachi, M., Satoh, J., Liaw, D. J.: Polym. Bulletin *1*, 581 (1979)
122. Lewis, F. M., Mayo, F. R., Hulse, W. F.: J. Am. Chem. Soc. *67*, 1701 (1945)
123. Mayo, F. R., Lewis, F. M., Walling, C.: J. Am. Chem. Soc. *70*, 1537 (1948)
124. Walling, C. et al.: J. Am. Chem. Soc. *71*, 1938 (1949)
125. Ham, G. E.: Copolymerization, pp. 1–87, New York: Interscience 1964
126. Alfrey, Jr., T., Bohrer, J. J., Mark, H.: Copolymerization pp. 45–73, New York: Interscience 1952
127. Alfrey, Jr., T., Overberger, C. G., Pinner, S. H.: J. Am. Chem. Soc. *75*, 4221 (1953)
128. Izumi, Z. et al.: J. Polym. Sci. *A3*, 2721 (1965)
129. Keber, R.: Makromol. Chem. *96*, 30 (1966)
130. Rybav, A. V., Semchikov, Y. D., Slavnitskaya, N. N.: Vysokomol. Soedin. *A12*, 553 (1970)
131. Zafar, M. M., Mohmud, R., Syed, M.: Makromol. Chem. *175*, 1531 (1974)
132. Saini, G., Leoni, A., Franco, S.: Makromol. Chem. *144*, 235 (1971)
133. Saini, G., Leoni, A., Franco, S.: Makromol. Chem. *146*, 165 (1971)
134. Saini, G., Leoni, A., Franco, S.: Makromol. Chem. *147*, 213 (1971)
135. Leoni, A., Franco, S., Saini, G.: Makromol. Chem. *165*, 97 (1973)
136. Jacob, M., Smets, G., De Schryver, F.: J. Polym. Sci. B *10*, 669 (1972)
137. Perec, L.: J. Polym. Sci. B *11*, 267 (1973)

138. Franco, S., Leoni, A.: Polymer *14*, 2 (1973)
139. Chatterjee, A. M., Burns, C. M.: Can. J. Chem. *49*, 3249 (1971)
140. Minsk, L. M., Kotlarchik, C., Darlak, R. S.: J. Polym. Sci., Polym. Chem. Ed. *11*, 353 (1973)
141. Pichot, C., Zaganiaris, E., Guyot, A.: J. Polym. Sci. C *52*, 55 (1975)
142. Chapiro, A.: Europ. Polym. J., *9*, 417 (1973)
143. Takemoto, K., Akashi, M., Inaki, Y.: J. Polym. Sci., Polym. Chem. Ed. *12*, 1861 (1974)
144. Akashi, M. et al.: J. Polym. Sci., Polym. Chem. Ed. *17*, 301 (1979)
145. Lee, M., Moriguchi, T., Minoura, Y.: J. Chem. Soc., Faraday I, *74*, 1738 (1978)
146. Lee, M., Utsumi, K., Minoura, Y.: J. Chem. Soc., Faraday I, *75*, 1821 (1979)

Received December 12, 1979
T. Saegusa (editor)

Kinetics of Polymerization Processes

Alexander Al. Berlin, Stanislav·A. Volfson, and Nikolai S. Enikolopian

Institute of Chemical Physics, USSR Academy of Sciences, Vorobyovskoye shosse 2b, USSR-Moscow B-334

Works pertaining to the following problem areas have been reviewed:
1) Radical polymerization, including the question of the dependence of chain termination rate constant on the length of the macroradical chain, the possibility for continuous radical polymerization to be achieved through the complexing and stabilization of free radicals and of catalytic chain transfer in radical polymerization.
2) Ionic polymerization, including the thermodynamic aspects of heterogeneous polymerization kinetics, kinetic features of the ionic polymerization of heterocyclic compounds and the zwitterionic polymerization.
3) Mathematical modeling of polymerization kinetics, including its application for the verification of kinetic schemes and the values of kinetic constants, as well as in the design of industrial processes.
    Investigations summed up in this review are mainly those carried out by kinetics schools in the Soviet Union over the past 5 years. The review contains 145 references to original works.

Table of Contents

List of Symbols 90

Introduction 92

A. **Free-Radical Polymerization** 92
    I. Dependence of Chain Termination Rate Constant on the Macroradical Chain Length 92
    II. "Living" Radical Polymerization 96
    III. "Living" Radical Copolymerization 98
    IV. Chain Termination Reactions in Radical Polymerization 101

B. **Ionic Polymerization Kinetics** 103
    I. Thermodynamic Aspects of Ionic Heterogeneous Polymerization Kinetics 103
    II. Polymerization Kinetics of Heterocycles 112
    III. Macrozwitterionic Polymerization 118

C. **Modeling the Kinetics of Polymerization Processes** 122
    I. Application of the Mathematical Simulation Methods to Verify the
       Kinetic Scheme and the Values of Kinetic Constants 122
    II. Investigations of Nonisothermal Polymerization Processes 133
    III. Problems of Designing Industrial Polymerization Reactors 136

References 137

List of Symbols

$[A]$:	concentration of active centres of polymerization
Ac:	acceptor monomer
a:	cross-section area of monomeric unit in polymer
C:	complex of comonomers
c:	heat capacity
C_p:	constant of chain transfer to polymer
C_m:	constant of chain transfer to monomer
C_s:	constant of chain transfer
D:	donor monomer
D_1:	diffusion constant of monomer
E:	activation energy
f:	initiation efficiency
i, j:	degree of polymerization
l, n:	length and number of segments in polymer zwitterion
k_{ij}:	termination rate constant of radicals R_i and R_j
k_t:	termination rate constant
k_p:	propagation rate constant
$k^0_{AA}, k^0_{AD}, k^0_{DD}$:	termination rate constants
k_s:	chain transfer rate constant
k_d:	depolymerization rate constant
k_i:	initiation rate constant
$k_{(-)}$:	rate constant of propagation on free ions
$k_{(\pm)}$:	rate constant of propagation on cyclic ion pairs
k_m:	rate constant of chain transfer to monomer
k_{tp}:	cross-termination rate constant
k_{tl}:	entangled radical termination rate constant
k_{AD}:	rate constant of copolymerization A and D
k_{AC}:	rate constant of copolymerization A and C
k_{DA}:	rate constant of copolymerization D and A
k_{DC}:	rate constant of copolymerization D and C
k_f:	rate constant of chain transfer to polymer
k'_d:	rate constant of deactivation
$K_p^{(n)}$:	equilibrium constant of ion pairs with degree of polymerization "n"
K_p:	equilibrium constant
$[M]$:	concentration of monomer
$[M]_{\lim}^{\text{liq}}$:	equilibrium concentration of monomer in relation to dissolved polymer
$[M]_{\lim}^{\text{sol}}$:	equilibrium concentration of monomer in relation to solid polymer
$[p]$:	polymer concentration
P_n:	number average degree of polymerization
P_w:	weight average degree of polymerization
q:	conversion of monomer
Q:	heat of reaction
$[R]$:	concentration of radicals
$[R_n]$:	concentration of "free" ions with degree of polymerization "n"
$[r_n]$:	concentration of ion pairs with degree of polymerization "n"
$[S]$:	concentration of chain transfer agent
$[T]$:	concentration of 1,3,5-trioxane

[Tr]:	concentration of 1,3,5,7-tetraoxane	μ_0:	chemical potential in standard state
V_p	molar volume of polymer	μ_{Mdiss}:	chemical potential of dissolved monomer
V_m:	molar volume of monomer		
V_s:	effective volume of ion pair	μ_{pdiss}:	chemical potential of dissolved polymer
W:	rate of reaction		
W_i:	rate of initiation	μ_{Tdiss}:	chemical potential of dissolved trioxane
W_{cr}:	rate of crystallization		
α:	$\dfrac{k_{AD}}{k_{DA}}$	$\mu_{TCpdiss}$:	chemical potential of trioxane units in the dissolved copolymer
ΔF:	change in free energy	ρ:	density
λ':	length of monomeric unit in polymer	η:	viscosity
		Φ_p:	volume fraction of polymer
λ:	heat conductivity coefficient		
δ:	Frank-Kamentsky's parameter	ν_{liq}:	average degree of polymerization of dissolved portion of polymer chain
μ:	kinematic viscosity		
μ_M:	chemical potential of monomer	τ:	induction period
		χ:	thermal diffusivity

Introduction

This review examines the research works, mainly of Soviet authors, on the kinetic laws underlying certain polymerization processes.

Recent years have ushered in new achievements in experimental techniques especially in the application of various physical methods in kinetic studies. Considerable changes have occurred in the approach to the investigation of kinetics. For example, formal kinetic analysis is now repeatedly combined with thorough chemical and physical studies of the object under investigation, with direct determination of intermediate compounds, complexes, etc.

The stock of kinetic techniques has considerably expanded as a result of the creation of effective methods for the analysis of molecular weight distribution (MWD) and the inclusion of MWD among the kinetic parameters. There is no longer a clearly defined line of demarcation between the free-radical and the catalytic polymerization.

Previously, experimental and calculational difficulties frequently forced investigators to confine themselves to studying the initial stages of polymerization, whereas now they deal more and more often with the whole process from beginning to end. This provides a logical foundation for modeling the kinetics and MWD of the whole process.

Naturally, all studies on the kinetics of polymer synthesis which have been published in recent years could not be included in this review. The authors have selected rather arbitrarily only those of direct interest to them.

The emphasis has been placed on the works of the Soviet school for the simple reason that Russian is still accessible only to a very limited circle of specialists all over the world.

This review is, to a considerable extent, a supplement to the authors' monograph published under the same title in 1978 by the "Khimiya" Publishers (Moscow).

A. Free-Radical Polymerization

I. Dependence of Chain Termination Rate Constant on the Macroradical Chain Length

The problem of quadratic termination and of the interaction between polymer radicals remains one of the most complicated questions in the kinetics of radical polymerization. The termination process is customarily regarded as a diffusion-controlled stage[1–3]. Indeed, in most cases the rate of chain termination in radical polymerization depends on the viscosity of the medium. This, among other considerations, explains the decreasing termination rate and the increasing polymerization rate in the course of the process, i. e., with the accumulation of the polymer in solution and the increasing diffusive hindrances which prevent the macroradicals from approaching and reacting with one another. In this case, according to numerous theories, where an attempt is made to describe quantitatively the termination reaction in radical polymerization, the recombination rate constant of radicals must depend on their degree of polymerization and decrease with the growing macroradical length following the $k \sim j^{-0.5}$ law (where j is the degree of polymerization). Dependencies of this type were actually observed in the processes of spin exchange[4], luminescence quenching[5], simulating the interaction of radicals, and in the recombination itself of ethylene oxide and styrene oligomers[6] in the range of degrees of polymerization from 100 to 1000.

What then are the kinetic corrollaries of such regular variations in the termination constant occurring in conjunction with the propagation of chain length? Assuming that the interaction rate constant for macroradicals whose degree of polymerization is i and j, can be expressed by $k_{ij} = k_0 \left(\dfrac{1}{i^{1/2}} + \dfrac{1}{j^{1/2}} \right)$, the increasing initiation rate in the processes where the mean degree of polymerization is determined by chain termination reactions will result in a decreased length of macromolecules and, therefore, in an increased effective termination rate constant. It is thus seen that polymerization rates increase in this case to a smaller extent than in the case of $k_{ij} =$ const. The observed order of the polymerization rate with respect to the initiation rate must be equal to 1/3, and not to 1/2, as in the case of $k_{ij} =$ const. It is to be noted that in a great majority of radical processes, the experimentally observed order with respect to the initiation rate is equal to 1/2. Another corollary of the dependence of termination rate on macroradical chain length is a broadening of the polymer molecular weight distribution, as compared with Flory's most probable distribution. This prediction, however, has not been experimentally confirmed either.

All this leads one to the conclusion that the termination rate constant in radical polymerization, at least for comparatively high degrees of polymerization (> 1000), does not depend on the length of the reacting macroradicals. Direct experimental data also seem to indicate this to some extent [4-7]. Apparently, the only possible explanation is that a decrease in the rate of progressive diffusion of macroradicals, when their length increases, is somewhat compensated by the growth of the effective cross-section of interaction due to the higher mobility of the end group of macromolecules, as compared with its center of gravity. This cross-section, however, does not seem to coincide with the macromolecule's effective radius, but is by about two orders of magnitude smaller and depends on the motion speed of the end group in the bundle, i.e., on the internal viscosity. It is otherwise impossible to explain quantitatively such small values of termination rate constants and the observed dependence between the interaction constant of the end groups of macromolecules and the length of the chain at degrees of polymerization < 100. An attempt to formulate a theory to deal with such a process of macromolecular interaction has been made in[8].

It should be noted that to date one more question related to the kinetics of the chain termination reaction in radical polymerization has not yet been conclusively solved. The diffusive nature of termination presupposes a quite definite, non-random distribution of radicals: the number of radicals located close to one another must be smaller than that which corresponds to the random law of particle distribution in space. The initiation reaction, i.e., that of the formation of radicals, does not depend on their location in space. This reaction will, therefore, distort the distribution of radicals in space that corresponds to the diffusive termination and will tend to make it random. This means that in a process using continuous initiation the termination rate constant must differ from that measured at post-polymerization. At continuous initiation the rate constant must be higher than the purely diffusive one.

Here we can draw an analogy with the equilibrium dissociation reaction, when the association rate constant in equilibrium is not limited by diffusion, regardless of the viscosity of the medium. In our opinion, this question requires at present a theoretical and experimental investigation. It is customary to assume that radical polymerization is characterized by a rather intensive chain termination reaction and a short time for the propagation of one chain, as compared to the time of polymerization. The existence of continuous processes ("living" polymers) has been ascertained for anionic[9] and cationic polymerization[10], where there is no bimolecular interaction of active centers with one another. Let us now examine certain radical polymerization processes in which the chain termination reactions are considerably inhibited or almost excluded.

For the rate of quadratic chain termination to be reduced, it is necessary to decrease sharply or arrest completely the progressive diffusion of the radicals. Among such processes are some heterogeneous polymerization reactions, when the polymer formed is not dissolved but separates in an individual, e.g. crystal phase, and where the mobility of occluded radicals is also not high. Similar effects are observed when three-dimensional polymers are formed from di- and polyfunctional monomers by the radical mechanism[11]. And, finally, intensive structurization of macromolecules in radical polymerization in the presence of complexing agents[12] also results in a decreasing termination rate and the formation of practically "living" polymers.

An interesting example is tetrafluoroethylene polymerization[13]. Poly(tetrafluoroethylene) is a highly crystalline polymer which is almost insoluble in any solvent. As a result of radical polymerization, or with irradiation of the polymer a considerable number of free radicals are formed and remain sufficiently stable when stored in a vacuum in the absence of the monomer. These radicals are detected by the ESR method and identified as chain propagation radicals. Brought into contact with the polymer which contains radicals, the monomer is polymerized. Moreover, during photoinitiated polymerization a prolonged post-polymerization period is observed for hours and even days after the light has been switched off. This indicates a considerable lifetime for propagation radicals in solid poly(tetrafluoroethylene). The post-polymerization rate, however, decreases with time and is possibly the result of either the annihilation (chemical transformation) of the radicals or their "immurement" in the polymer formed (physical deactivation). Kinetically, the second process, as well as the first can be described by the law of the second order reaction, since the termination rate must be proportional to the concentration of radicals and the rate of polymer formation. This rate is also proportional to the concentration of radicals:

$$W_t \sim [R] \frac{d[M]}{dt} \sim [R] k_p \cdot [R][M] \sim k_p [M][R]^2 \qquad (1)$$

As seen from Eq. (1), the rate of decrease in the concentration of active radicals must, in this case, also be proportional to the monomer concentration. The post-polymerization kinetics was indeed satisfactorily described, as will be seen later, proceeding

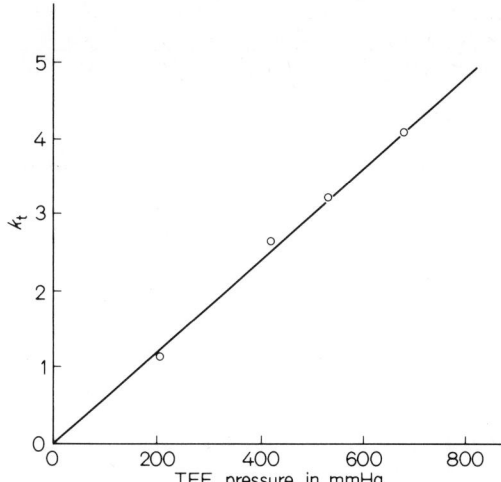

Fig. 1. Dependence of the effective termination rate constant on tetrafluoroethylene (TFE) concentration; k_t in arbitrary units

from this assumption. Using the ESR method has shown, however, that in actual fact the absolute number of radicals also decreases thereby causing a proportional decrease in the reaction rate. Thus, it is evident that in this system the radicals are annihilated in the presence of the monomer, and also that the rate of annihilation is proportional to the monomer concentration (Fig.1; the results shown in Fig. 1 were kindly provided by the authors of paper[13]).

In order for the chemical interaction to take place, radicals must come into contact with one another. The dependence of the rate of this reaction on the monomer concentration makes it possible to assume that the transfer of radicals in space is brought about by the propagation reaction, i.e., by the addition of monomer molecules. Assuming that such a transfer in space is of a random nature from the point of view of the radicals approaching each other, and that the radicals react instantaneously at a distance r from each other (the reaction is limited by transfer in space, and not by the chemical interaction), the authors of[13] obtained the following equation to describe for the concentration of radicals (R) and the yield of polymer (q):

$$-\frac{d[R]}{dt} = \kappa [R]^2$$

$$q = q_0 (1 + \kappa [R]_0 t)^\beta$$

where $\beta = \left(1 + \frac{4\pi\lambda r}{a}\right)^{-1}$, $\kappa = k_p [M] \lambda a \left(1 + \frac{4\pi\lambda r}{a}\right)$

k_p is the propagation rate constant, [M] is the monomer concentration, and λ and a are the length and cross-section of the monomeric unit in the polymer. Experimental results were adequately described by the proposed reaction scheme (Fig. 2), and the radius had an acceptable value (1 – 4 Å).

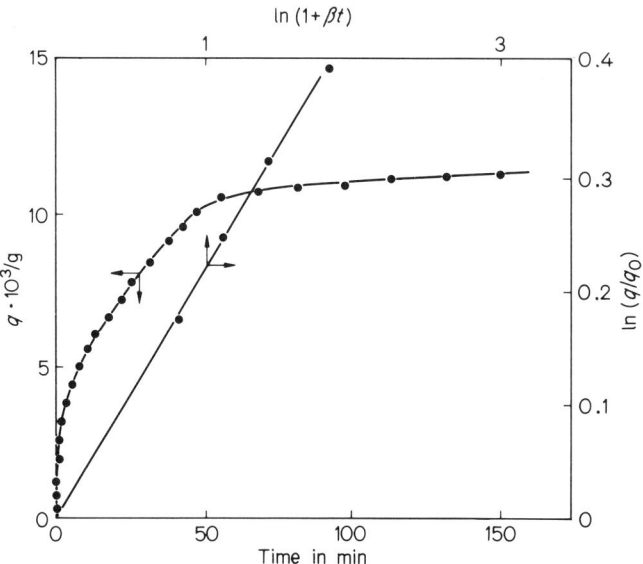

Fig. 2. Kinetics of TFE polymerization on γ-irradiated poly(tetrafluoroethylene) (polymer sample weight $q_0 = 0.01$ g, $[R] = 4.5 \cdot 10^{17}$ cm^{-3}; TFE pressure 418 mm Hg; $T = 102\,°C$)

II. "Living" Radical Polymerization

It was recently shown[12] that in radical polymerization the chain termination rate constant is observed to decrease with the introduction of a polyfunctional complexing agent into the system. An especially sharp decrease of the termination rate, up to the formation, under certain conditions, of "living" radical polymerization centers, was noted in the methyl methacrylate-orthophosphoric acid system.

At continuous photoinitiation the methyl methacrylate polymerization rate was shown to increase considerably with the introduction of H_3PO_4; the reaction is inhibited by a typical radical inhibitor – p-benzoquinone, and the values of the propagation rate constant are similar both with and without the presence of H_3PO_4. Everything indicates that this is a radical process, and that the role of the complexing agent is essentially to reduce the chain termination rate. Under conditions of continuous initiation the observed order with respect to the initiation rate of radicals is close to 0.5 (0.6); i.e., most radicals are still annihilated in the course of their bimolecular interaction. If, however, the post-polymerization is studied with the light off, it turns out that the polymerization rate is at first somewhat decreased; then, for a long time it remains constant, with the yield reaching 100%. The molecular weight of the polymer grows, while the number of chains remains almost constant. The introduction of benzoquinone reduces the post-polymerization rate and, at a certain concentration, completely inhibits the process. It is interesting to note that a similar effect is also produced by a chain transfer agent (1-butenethiol): with the growth of its concentration both the molecular weight and the reaction rate decrease. All these data, as well as the laws underlying the variations of polymer molecular weight distribution, are logically explained by the assumption that, with the growing degree of macroradical polymerization, the probability increases of its being entrapped in a certain associate as a result of complexing. This reduces its mobility and, consequently, decreases the rate constant of its interaction with other radicals. While continuous initiation exists, termination is actualized through the interaction of low-molecular-weight radicals with high-molecular weight radicals. After all the radicals have reached a certain length, the chain termination practically stops. In[12] a model is proposed which takes into account the dependence of the termination rate constant on chain length; when the form and parameters of this dependence are adequately chosen, this model satisfactorily describes the experimental data.

The effect produced by the complexing agent in the liquid phase was observed at relatively high temperatures (close to room). Here the trifunctionality of the complexing agent seems to be the controlling factor.

A similar effect was observed under different conditions at low temperatures in high-viscosity systems. In[14] the authors studied post-polymerization of butyl methacrylate and other acrylates, which were γ-irradiated in glassy state with subsequent heating at a constant rate. The action of inhibitors, such as benzoquinone and 2-methyl-2-nitrosopropane, is evidence of the radical mechanism of the reaction. Variation of the concentration of radicals was studied using the ESR method. It was shown that at first, when the temperature is low, the primary radicals which have monomer molecules attached to them are transformed into propagation radicals; then, in the glass transition region the radicals are partially annihilated. The kinetics of the annihilation of radicals depends on the defreezing rate. At sufficiently slow defreezing rates (Fig. 3), a certain portion of the radicals is stabilized and not annihilated when raised to room temperature, polymerization of these radicals correspondingly proceeding according to the "living" polymer type. The annihilation or pre-

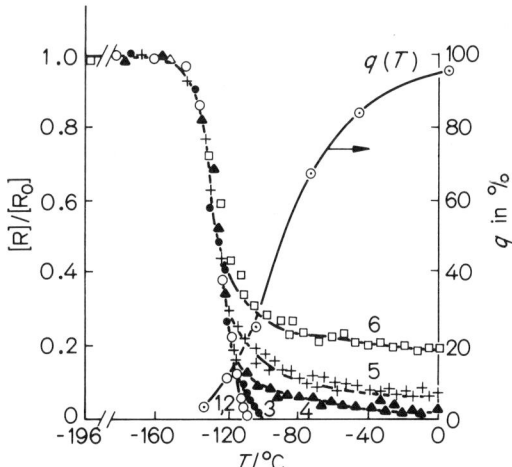

Fig. 3. Change of radical concentration in butyl methacrylate, γ-irradiated at $-196\,°C$. Heating rate °C/h: 1, 60; 2, 30; 3, 9; 4, 6; 5, 4,8; 6, 3. $q\,(T)$ – polymer yield during heating at 5 °C/h

servation of active radicals in the course of sample heating is determined by competing recombination and chain propagation reactions, as a result of which the macroradical length increases and, correspondingly, the mobility and the probability of recombination decrease. Based on this reasoning, it can be assumed that by defreezing of the monomer with the lowest temperature region of vitrification 2-allyloxyethyl methacrylate (allylmethacrylic ether of ethylene glycol, between $-169°$ and $-167\,°C$ (?)), the annihilation reaction prevails, whereas by defreezing the monomer with the highest temperature region of vitrification 2-hydroxyethyl methacrylate (monomethacrylic ether of ethylene glycol, between -94 and $-88\,°C$), even at the highest heating rate ($\approx 60\,°C/h$), about half of the propagation radicals are preserved, and the polymer yield reaches 100%. All other things being equal (the temperature range of glass transition and the yield of primary radicals are in both cases practically the same), the addition of inert low-molecular-weight diluent (isobutyl butanoate) to the monomer decreases the chain propagation rate and increases the mobility of macroradicals. For this reason the polymer yield is reduced with the introduction of such a diluent. On the other hand, the introduction of poly(butyl methacrylate) into the initial monomer increases the viscosity of the system and raises the probability of macroradical stabilization in the course of propagation, thereby increasing the yield of the "living" radicals and the polymer yield. In this case also, the insertion of complexing agents (e.g., $ZnCl_2$) into the monomer increases the predisposition of the formed polymer and macroradicals to become structured while the percentage of "living" radicals and the polymer yield increase.

III. "Living" Radical Copolymerization

In studies[15, 16] systems such as methyl methacrylate (MMA)/H_3PO_4, acrylonitrile (AN)/H_3PO_4, and methyl α-chloroacrylate (MCA)/H_3PO_4 have been used to produce block copolymers, graft polymers, and graft block copolymers on solid organic (cellophane film) and inorganic (silica gel) substrates. The method applied was the one previously developed[9] for ionic systems: the successive introduction of new portions of comonomers. At first it was shown that after all the monomer had been consumed in the H_3PO_4 : MMA = 3 : 1 (mole ratio) system in post-polymerization (initiation with UV light) with a new portion of methyl methacrylate introduced the amount of the polymer increases, along with a proportional rise in the molecular weight. The introduction of another monomer in the second stage results in the formation of two fractions: a block copolymer with a higher molecular weight formed on the living radicals, and a homopolymer – the "dead" polymer of the first stage of the process. It should be noted that a homopolymer is never formed in the second stage of the process; i.e., the "living" radicals initiate the polymerization of the second portion of the monomer exclusively through addition. Of special interest is the possibility of producing a considerable amount of graft polymers on solid substrates. Acrylic monomers (MMA or MCA), along with orthophosphoric acid, were applied to previously γ-irradiated cellophane film or silica gel in vacuum. A graft polymer was observed to form, but almost no homopolymer was formed in this case. Without a complexing agent, polymerization does not take place under the same conditions.

It seems that in this case as well, in the absence of a structure-forming agent the mobility of the radicals on the surface is sufficiently high; in addition they rapidly recombine. Successive introductions of portions of different monomers made it possible to produce double and triple block copolymers, e.g., PMMA-PMCA-PMMA or PMCA-PMMA-PMCA.

"Living" radical processes thus open up new possibilities of synthesis, previously characteristic of ionic "living" systems: the production of block copolymers, and the established possibility, in principle, of obtaining copolymers with a narrow molecular weight distribution. In studies[15, 16] the formation of "living" radical systems was accompanied by a partial annihilation of radicals, i.e., the formation of a "dead" polymer, which corresponded to the broadening of MWD. However, there seem to be no fundamental difficulties in producing polymers with a narrow MWD through the use of a similar method.

The formation of alternating copolymers through the polymerization of pairs of monomers, one of which is the donor and the other the acceptor of an electron, is well known. We shall mention only a few studies out of a great number of those recently published. First, those dealing with the nature of active centers in such systems will be examined. When radical initiators are used, e.g., benzoyl peroxide as in[17], and the reaction is inhibited with different radical polymerization inhibitors, such as stable radicals like 2,2,6,6-tetramethylpiperidine 1-oxide, quinones, fluorene etc., questions concerning the nature of active centers can be regarded as solved. In[17, 18] ESR was used to investigate the structure of radicals formed with γ- and UV irradiation of the mixtures of 2,3-dimethylbutadiene (DMBD) with SO_2, vinyl acetate with SO_2, and maleic anhydride with 2,3-dimethylbutadiene. It was shown that in the 2,3-dimethylbutadiene/SO_2 system at UV irradiation (at $-196\,°C$), only $\sim SO_2^{\cdot}$ propagation radicals are formed, at γ-irradiation (also at $-196\,°C$) $\sim SO_2^{\cdot}$ radicals and DMBD˙ allyl radicals are formed. With heating of the sample in the range of the glass transition temperature, intensive polymerization takes place, and an alternating copolymer of 50:50 (mole ratio) composition is formed; in the ESR spectrum of γ-irradiated samples, both types of propagation radicals are observed, but in the UV radiated, only $\sim SO_2^{\cdot}$ is found. Moreover, the introduction of a strong donor solvent (dimethylformamide) results in a decrease in the activity of the $\sim SO_2^{\cdot}$ radical and a practically complete suppression of polymerization in UV irradiated samples. At the same time, when γ-irradiated samples are heated, a substantial amount of the alternating copolymer is formed. These data indicate that in this system (similar results were also obtained for the vinyl acetate/SO_2 system) the formation of an alternating copolymer is caused not by the successive cross-addition of monomers (with the values of copolymerization constants $r_1 \ll 1$, $r_2 \ll 1$), but by independent polymerization of a complex of monomers both on the $\sim SO_2^{\cdot}$ and the $\sim$DMBD˙ radicals, with each of the radicals actively participating in the copolymerization process.

Another result was obtained for the systems in which the acceptor is not SO_2 but maleic anhydride[17, 18]. These authors investigated the polymerization of 2,3-dimethylbutadiene/maleic anhydride (MA) mixture γ-irradiated at $-196\,°C$. With the heating of irradiated samples, post-polymerization is observed at glass transition, and an alternating copolymer of 50:50 composition is formed. Using ESR the authors observed that the overall concentration of propagation radicals decreased and, that against this background, the absolute number of $\sim$ DMBD· radials grew. This means that 2,3-dimethylbutadiene molecules are being attached to the $\sim$ MA· radical. In this case, therefore, polymerization is effected by the successive addition of monomer molecules to propagation radicals. A similar result was obtained in[19] using a kinetic method for the indene/maleic anhydride system. Using spectral methods the authors studied the polymerization initiated with benzoyl peroxide, as well as the complexing of monomers. The following possible chain propagation reactions were examined:

$$\sim Ac· + D \rightarrow \sim D· \quad k_{AD}$$
$$\sim Ac· + C \rightarrow \sim Ac· \quad k_{AC}$$
$$\sim D· + Ac \rightarrow \sim Ac· \quad k_{DA}$$
$$\sim D· + C \rightarrow \sim D· \quad k_{DC}$$

where Ac is the acceptor monomer (e.g. maleic anhydride), D is the donor monomer (e.g. indene), and C is the complex of comonomers. The equation for the reduced reaction rate has the following form:

$$\frac{W}{[Ac]} = F(\varphi) K_p \left(\frac{k_{AC}}{k_{AD}} + \frac{k_{DC}}{k_{DA}} \varphi \right) [Ac] + 2 F(\varphi)$$

where K_p is the complexing equilibrium constant,

$$F(\varphi) = \frac{W_I^{1/2} \cdot k_{AD} \cdot \varphi}{(k_{AA}^0 + 2 k_{AD}^0 \alpha \cdot \varphi k_{DD}^0 \alpha^2 \varphi^2)^{1/2}},$$

k is the termination rate constant, and $\varphi = \frac{[D]}{[Ac]}$. The reduced polymerization rate vs. the maleic anhydride [Ac] concentration curves proved to be parallel to the abscissa for different values of the parameter φ. This means that $k_{AC} \ll k_{AD}$ and $k_{DC} \ll k_{DA}$, i.e., the successive cross-reactions of monomer addition in the given system, are the main reactions of polymer formation.

The following question arises: why is it that the condition of $k_{11} \ll k_{12}$ and $k_{22} \ll k_{21}$ is met in these systems? Complexing seems to play a significant part here, not between the monomers, but between donor radical and acceptor monomer and acceptor radical and donor monomer[19]. This assumption is quite natural; what is needed now is proof to support this hypothesis.

It can thus be regarded as established that the formation of alternating copolymers in these systems proceeds in accordance with the radical mechanism; also, depending on the acceptor monomer, both the addition of the molecules of the complex (SO_2) and successive additions of molecules of the monomer (maleic anhydride) can be considered the main reactions. In the latter case, the complexing of radicals with monomers seems to be the controlling factor in the formation of a sequence copolymer.

IV. Chain Termination Reactions in Radical Polymerization

It was initially thought that the mechanism of free-radical polymerization was comprised of the stages of initiation, propagation, and termination of the chain. In 1937 Flory[20] introduced the concept of a chain transfer consisting of two elementary acts:

the chain transfer proper to the molecule of the substrate S:

$$R_n^{\cdot} + S \rightarrow P_n + S^{\cdot}$$

and the chain regeneration proceeding at a high rate:

$$S^{\cdot} + M \rightarrow R_1^{\cdot} + S$$

Thus, the limiting reaction is always the first one.

Later, in the studies of Breitenbrach[21], Medvedev[22], and Mayo[23], it was shown that in a great majority of cases the action of chain transfer consists of the ejection of a hydrogen or halogen atom from the substrate molecule:

$$R_n^{\cdot} + R'H \rightarrow R_nH + R'^{\cdot}$$
$$R_n^{\cdot} + CCl_4 \rightarrow R_nCl + CCl_3^{\cdot}$$

There also exists the possibility of chain transfer with the splitting-off of a whole group[24] as in the following:

$$R_n^{\cdot} + R'-S-S-R'' \rightarrow R_n-S-R' + R''S^{\cdot}$$

A specific case of chain transfer is the transfer to the monomer which limits the magnitude of molecular weight at a low initiation rate and at elevated temperatures. This elementary action can have two possible mechanisms:

(a) the radical ejecting the hydrogen atom from the monomer molecule:

$$\sim CH_2 - \overset{\cdot}{C}H + CH_2 = CH \rightarrow \sim CH_2 - CH_2 + CH_2 = \overset{\cdot}{C}$$
$$\quad\quad\quad | \quad\quad\quad\quad | \quad\quad\quad\quad\quad\quad | \quad\quad\quad\quad\quad | $$
$$\quad\quad\quad X \quad\quad\quad\quad X \quad\quad\quad\quad\quad\quad X \quad\quad\quad\quad\quad X$$

(b) a disproportionation according to the scheme:

$$\sim CH_2 - \overset{\cdot}{C}H + CH_2 = CH \rightarrow \sim CH = CH + CH_3 - \overset{\cdot}{C}H$$
$$\quad\quad\quad | \quad\quad\quad\quad | \quad\quad\quad\quad\quad\quad | \quad\quad\quad\quad\quad | $$
$$\quad\quad\quad X \quad\quad\quad\quad X \quad\quad\quad\quad\quad\quad X \quad\quad\quad\quad\quad X$$

Because of the low magnitude of the chain transfer constant for chain transfer to the monomer for all vinyl groups, it has until now been impossible to establish the nature of the end groups and make a choice in favor of one or the other of the mechanisms.

Characteristic values for chain transfer to different compounds are usually cited in reference books.

The dependence of the average degree of polymerization on the concentration and activity of the chain transfer agent is described by Mayo's equation:

$$\frac{1}{\bar{P}_n} = \frac{1}{\bar{P}_n(\infty)} + C_s \frac{[S]}{[M]}$$

where $C_s = k_s/k_p$.

The higher the activity of the chain transfer agent, the more it reduces the average degree of polymerization, but the faster it is used up in the course of the process:

$$\frac{[S]_t}{[S]_0} = \left(\frac{[M]_t}{[M]_0}\right)^{C_s}$$

This results in a falling-off of the chain transfer rate in the course of the process, thereby causing a substantial change in molecular weight and molecular weight distribution of the products. This makes it necessary for technologists to use as chain length regulators such substances whose relative chain propagation constant is close to unity. Only in this case is a uniform consumption of monomer and regulator achieved.

The concentration of these regulators must obviously be relatively high, although this is not always desirable.

A special case of chain transfer is that of the chain to the polymer with termination, discovered by N. S. Enikolopian in 1960[25] while observing the polymerization of heteroatomic compounds

$$\sim C^+ + \underset{\underset{\{}{C}}{\overset{\overset{\}}{C}}{O}} \longrightarrow \sim CO + C^+$$

As a result of this reaction, the number average degree of polymerization does not change, although the weight average degree of polymerization changes in such a way that, regardless of the initial MWD of the product, it tends to approach Flory's distribution. This reaction has been applied to the modification of end groups and to the production of statistical block copolymers.

Another question to be considered is whether it is in principle possible to affect the course of radical polymerization catalytically. The ions of metals with a variable valence are well known to be capable of catalyzing the monomolecular decomposition of organic peroxides and hydroperoxides.

Of special interest is the possibility of a direct effect on the magnitude of the rate constant of chain transfer to the monomer and the chain termination rate constant.

After much research a group of investigators[26] succeeded in finding a number of selective complexing agents and proving the possibility of selective catalysis for the two stages of radical polymerization.

The porphyrin cobalt complex in radical polymerization of methylmethacrylate catalyzes the chain transfer to the monomer without affecting the polymerization rate. The phthalocyanine cobalt complex catalyzes the chain termination.

When MMA was subjected to radical polymerization in block, the initial polymerization rate decreased by no more than 20% with the concentration of the complex equal to $1 \cdot 10^{-3}$ mol/l. The order of the reaction with respect to the initiator was preserved, but the gel effect practically disappeared, in addition the ultimate conversion increased to 100%, with the concentration of the complex equal to $7 \cdot 10^{-4}$ mol/l.

Throughout, the reaction rate is described by the classic equation

$$W = k_p k_t^{-0.5} k_I^{0.5} [M] [I]^{0.5}$$

The number of polymer chains formed as a result of polymerization is by 3 orders of magnitude higher than that of the complexing agent molecules entering the reaction and by 2 orders of magnitude higher than the number of initiator molecules.

The degree of product polymerization was defined by Mayo's equation and did not depend on temperature ($C_s = 2.4 \times 10^3$). The absolute chain transfer constant $k_s = 1.4 \times 10^6$ $mol^{-1} s^{-1}$.

This value is in agreement with the rate constants of chain termination by metal complexes measured by Bamford[27] and Dainton[28].

B. Ionic Polymerization Kinetics

The following questions are discussed in this section:
1) thermodynamic aspects of heterogeneous polymerization kinetics, as exemplified by ionic processes;
2) kinetic features of the ionic polymerization of heteroatomic compounds, including nitrogen-containing cycles, ethylene oxide, cyclic acetals, and formaldehyde;
3) characteristic features of the macrozwitterionic polymerization of vinyl monomers.

I. Thermodynamic Aspects of Ionic Heterogeneous Polymerization Kinetics

Homogeneous reversible polymerization is known to be characterized by a limiting equilibrium concentration of the monomer, below which a high-molecular-weight polymer is not formed, and, correspondingly, by a limiting temperature, above which (at $\Delta H° < 0$, $\Delta S° < 0$) it is also impossible for a high-molecular-weight polymer to form[29]. The situation is more complicated when the polymer is isolated in a separate phase[30].

When studying the cationic polymerization of 1,3,5-trioxane (the short form "trioxane")[31, 32], investigators ran into peculiarities of formal kinetics, such as apparently high orders with respect to the monomer and their relationship to the nature of the solvent, temperature, etc., induction periods that could not in any way be ex-

plained by the results obtained from contemporary research on ionic liquid-phase processes. In papers[33, 34] a hypothesis was formulated to explain all the peculiarities of the process in various solvents while taking into account three characteristic features of trioxane polymerization:
1) Reversibility of the process;
2) Heterogeneous nature of the system (the polymer dissolves neither in the monomer nor in the solvents used for polymerization);
3) Formation of polymerizable by-products (formaldehyde, 1,3,5,7-tetraoxane, 1,3,5,7,9-pentaoxane, and other cycles).

The polymerization system and the processes taking place in it can be presented as follows: The growing macromolecule can be either fully crystallized so that the active center is located on the surface of the crystal (crystalline polymer); or it can be partially dissolved in the medium, with the active center located at the dissolved end of the chain section (dissolved polymer). The equilibrium monomer concentrations of the polymer differ in the dissolved and the crystalline state. Addition to the surface active center with the formation of a solid polymer is thermodynamically more favourable than addition to the active center in solution with the formation of a dissolved polymer.

The processes taking place in the polymerization system are: reversible processes of the monomer while being attached to the active center at the dissolved end section and the reverse process of depolymerization of this chain section; chain propagation and depolymerization at the active centers on the crystal surface; and crystallization and dissolution of the propagating chain. The nature of the dependence of reaction rate on monomer concentration is determined by the ratio between the rates of these processes. If the crystallization rate is slower than the propagation rate of the dissolved portion of the macromolecule, chain propagation takes place mainly on the active centers at the ends of the dissolved parts of macromolecules. The process is described by the equation of reversible polymerization with an equilibrium monomer concentration corresponding to the monomer − dissolved polymer equilibrium (Fig. 4, straight line 1).

$$W = k_p [A] ([M] - [M]_{\lim}^{liq}) \qquad (2)$$

Free energy variation at the conversion of dissolved trioxane into dissolved polymer, as in the case of other six-membered rings, is not great; i.e., the reaction is reversible, and the equilibrium concentration of the monomer is fairly high. When the monomer concentration decreases, the propagation rate of the dissolved part of the polymer molecule becomes slower than the crystallization rate; the kinetics is described by the equation of reversible polymerization with an equilibrium concentration in relation to the crystalline polymer (Fig. 4, straight line 2). It must be noted that at monomer concentrations lower than $[M]_{\lim}^{liq}$, the formation of long sections of polymer chain in solution is thermodynamically unfavorable. Therefore, only a short oligomer section of the chain exists in solution. This section is always in equilibrium with the monomer, whereas the formation of a high-molecular-weight polymer is actualized by a slow crystallization of this oligomer section, with the reaction rate equal to

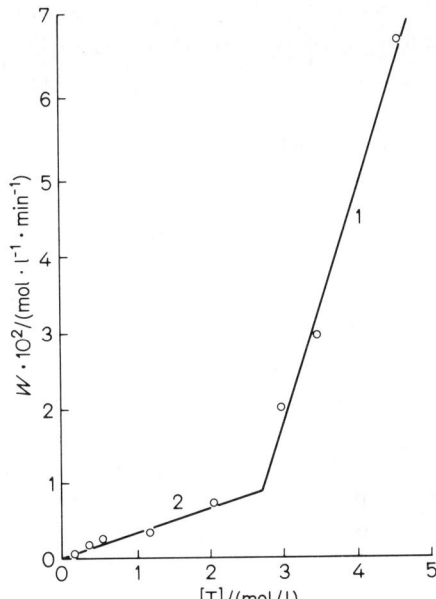

Fig. 4. Dependence of the maximum polymerization rate of trioxane in methylene dichloride on the initial monomer concentration. (P_{diss} – dissolved polymer, P_{solid} – solid polymer)

$$W = \frac{W_{cr}}{k_d} k_p ([M] - [M]_{lim}^{solid}) \; A = k \, [A] \, ([M] - [M]_{lim}^{solid})$$

Extrapolation of the linear portions of the curve in Fig. 4 to $W = 0$ yields values of equilibrium concentrations of trioxane in methylene dichloride at 20 °C; 2.5 mol/l (Fig. 4, straight line 1) for monomer – dissolved polymer equilibrium and 0.15 mol/l for monomer/crystalline polymer equilibrium (line 2).

The value of equilibrium concentration must not depend on the type and the concentration of the catalyst; this has been observed experimentally for the catalysts $SnCl_4$, $BF_3 \cdot O(C_2H_5)$, and $(C_2H_5)O^+ \, SbCl_6^-$.

The value of equilibrium concentration of the monomer in relation to the dissolved polymer, as estimated from the dependence of polymerization rate on the initial monomer concentration, correlates well with the value of this concentration measured by independent methods[35].

With increasing temperatures the value of equilibrium concentration must rise, a finding shown by the measurement of $W - [M]_0$ dependence at different temperatures in two solvents: nitrobenzene and chlorobenzene. From these data the heats of dissolved monomer – dissolved polymer transition were estimated in[34, 36, 37].

Valuable information on the mechanism of the process and on the confirmation of the formulated assumptions was obtained by analyzing the low-molecular-weight by-products of trioxane polymerization reaction: 1,3,5,7-tetraoxane and formaldehyde. Theoretical analysis has shown that, depending on the state of active centers (surface or dissolved) and the length of the dissolved portion of the polymer chain, the steady-state concentration of 1,3,5,7-tetraoxane and formaldehyde changes. A comparison between experimental and theoretical data has shown that at monomer concentrations $[M] > [M]_{lim}^{liq}$, the active centers are in a dissolved state. This process can be described by the following scheme:

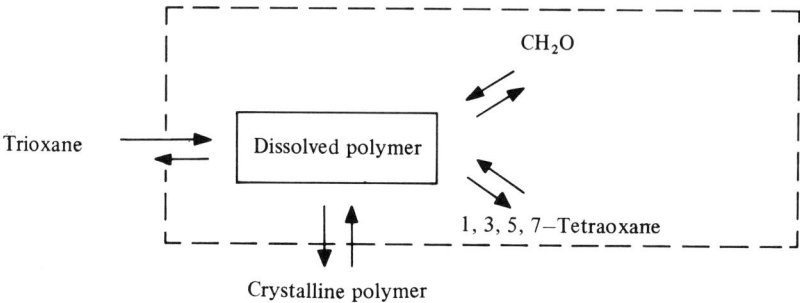

The dashed line encloses the compounds that are in equilibrium with one another. Formaldehyde concentrations in this case are equal to the limiting[a] concentrations with respect to the dissolved polymer and do not depend on the concentrations of trioxane. At monomer concentrations $[M] < [M]_{lim}^{liq}$, part of the active centers is in solution and the other is on the surface; the average length of the dissolved portion of the polymer chain depends on the monomer concentration in accordance with an equation similar to Tobolsky's

$$\gamma_{liq} = \left(1 - \sqrt[3]{[T]/[T]_{lim}^{liq}}\right)^{-1}$$

The process is described by the scheme:

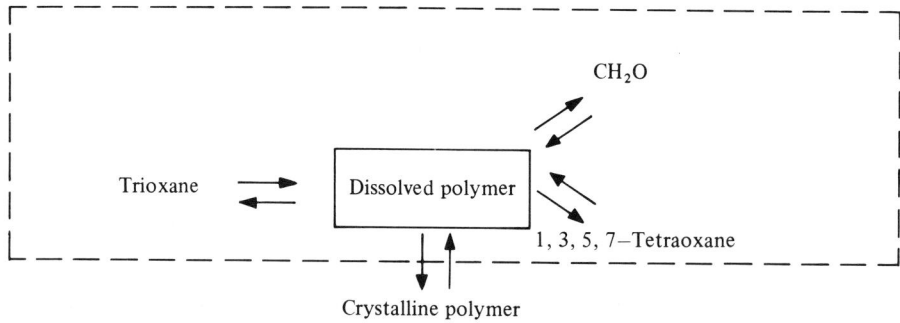

Formaldehyde, trioxane, 1,3,5,7-tetraoxane, and dissolved oligomer chains (with a small degree of polymerization) are here in equilibrium. Therefore, the steady-state concentrations of formaldehyde and 1,3,5,7-tetraoxane depend on trioxane concen-

---

[a] By limiting concentration we understand the equilibrium concentration with respect to an infinitely long molecule

tration: $[Tr] \sim [T]^{4/3}$. The point of inflexion on the curve which describes the dependence 1,3,5,7-tetraoxane steady-state concentration on the concentration of trioxane corresponds to the trioxane — dissolved polymer of infinite length — 1,3,5,7-tetraoxane equilibrium. The values of $[Tr]_{lim}^{liq}$ in nitrobenzene and methylene chloride which were found in this way are in good agreement with $[Tr]_{lim}^{liq}$ values determined from $W - [M]_0$ curves and by other independent methods.

The ideas presented above also made it possible to explain some other features of trioxane polymerization kinetics (the dependence of induction period on monomer concentration, the "limitedness" of kinetic curve, etc.). Moreover, in accordance with the same ideas, the rate vs. monomer concentration curve can pass through a maximum (at $[M] < [M]_{lim}^{liq}$) if the active centers on the surface of polymer crystal are more reactive than the "dissolved" active centers. The extreme dependence of the polymerization rate on monomer concentration

$$W = k_p [A] \left(1 - \frac{[M]}{[M]_{lim}^{liq}}\right) ([M] - [M]_{lim}^{sol}) \quad (\text{at } [M]_{lim}^{sol} < [M] < [M]_{lim}^{liq})$$

is explained by the finding that the concentration of surface active centers decreases with increasing monomer concentration until at $[M] > [M]_{lim}^{liq}$ all the active centers pass into solution, at which time the rate dependence on monomer concentration is described by Eq. (2). It is thus seen that polymerization rate at first increases, passes through a maximum, then through a minimum, and increases again as monomer concentration increases.

Such complicated dependencies are actually observed in the polymerization of trioxane in cyclohexane and heptane[34]. Similar catalytic phenomena have also been observed in reversible heterogeneous polymerization of 1,4-diazabicyclo[2.2.2]-octane another cyclic monomer[38].

One of the objectives of the investigators during their research into the thermodynamic characteristics of the formaldehyde — trioxane — 1,3,5,7-tetraoxane — ... — dissolved poly(oxymethylene) — crystalline poly(oxymethylene) system, was to verify the assumptions concerning the mechanism of trioxane polymerization stemming from formal kinetic analyses. These investigations are also of interest in the elaboration of polymerization thermodynamics, and, in particular, the polymerization of heterocycles. Much attention has been devoted to ascertaining the environmental effects of the solvent, polymer, etc., on polymerization thermodynamics[36, 39].

Among other subjects the influence of the conditions, in which the polymerization process is conducted, on the supramolecular structure of the polymer formed is discussed in studies[40–42]. Using these data the researchers analyzed the impact of polymerization kinetics on the polymers supramolecular structure and formulated the basic principles for controlling the structure of the polymer in the course of its synthesis. They also proposed a new thermodynamic approach for controlling supramolecular structures. The possible uses of this method are demonstrated in the polymerization of trioxane and triethylamine in different solvents and at different monomer concentrations. The purpose of this approach and the manner in which it differs from the conventional "kinetic" approach are roughly illustrated by the scheme in Fig. 5.

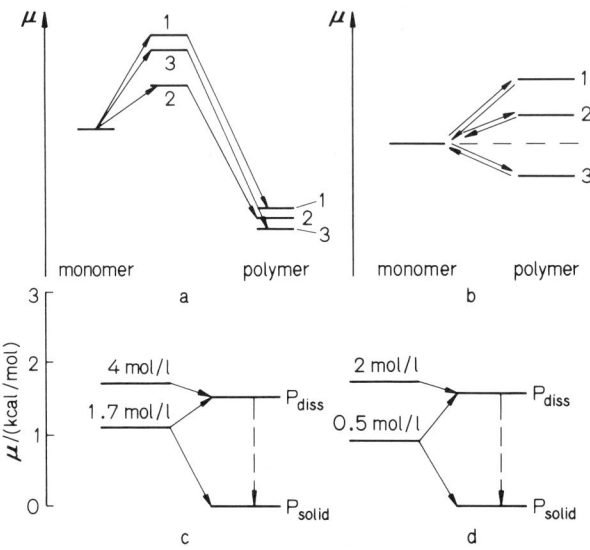

Fig. 5 a–d. Energy diagrams for irreversible (**a**) and reversible (**b**) processes of polymer formation in trioxane, polymerization in methylene dichloride (**c**) and nitrobenzene (**d**); (1, 2, 3: different polymer structures)

Polymerization reactions are in most cases conducted under conditions remote from polymer-monomer equilibrium, i.e., these reactions are, to a great extent, irreversible. A polymer can have different molecular and supramolecular structures; for example it can be iso- or syndiotactic, amorphous, or crystalline. Differences in the chemical potential of polymers with different structures are small in comparison with the changes observed in the chemical potential at conversion of a monomer into a polymer. This means that the possibility for a polymer of a certain structure to be formed will be determined by kinetic causes: the nature of the catalyst, solvent, etc. According to the scheme in Fig. 5, the polymer with structure 2 will be mainly produced.

If, after the polymer has been formed, a transformation of one structure into another is possible (e.g., formation of an amorphous polymer with its subsequent crystallization), the kinetic characteristics of these transformations will, in their turn, exert the determining effect on the final structure of the polymer. Specifically, the supramolecular structure of a polymer produced in the course of its synthesis will change, depending on the relationship between the rates of three processes: (1) chemical reaction of polymer formation, (2) isolation of polymer in a separate phase, (3) structural transformations inside the polymer phase. In the latter two processes, a significant role is played by the ratio between the rates of the formation and growth of the nuclei of one phase inside the other. This is the kinetic aspect of the problem of controlling the polymer structures during synthesis.

What is then the basis for the thermodynamic approach? One can select the reaction conditions in such a way that variations in the chemical potential during the process will be comparable to the difference in the chemical potentials of polymers with a different structure. Thus, it is impossible to conduct the reaction under conditions closely approximating the equilibrium of a number of structures (according to the scheme in Fig. 5, the first and the second) for thermodynamic reasons, and the polymer formed will have structure 3 (according to the scheme in Fig. 5).

For purposes of analogy we can cite the process of growing monocrystals of low-molecular-weight substances. To grow a perfect crystal, it is necessary to keep the crystallization conditions (temperature and pressure) very similar to the conditions of gas — crystal or liquid — crystal equilibrium. This is done to avoid the formation of new solid-phase nuclei and significant defects in the crystal, since both require additional energy. The scheme in Fig. 5 shows the chemical potential of the liquid (or gas) on the left, the chemical potential of the monocrystal (3) on the lower right, and the chemical potential of the critical solid phase nuclei and crystal defects above (2 and 1). How then can one vary the values of the chemical potentials of the polymer in a different structure and the monomer? The chemical potentials of polymers and monomers do not usually depend to the same extent on the temperature, pressure and nature of the solvent. Moreover, the chemical potential of a monomer (μ_M) depends significantly on its concentration in solution (or partial pressure in the gas phase). To a first approximation, this dependence on an ideal solution has the form:

$$\mu_M = \mu_0 + RT \ln [M]$$

Thus, by varying the conditions in which polymerization is conducted (temperature, pressure, nature of solvent and/or concentration of the monomer), one can select and control the polymer structure.

To confirm the correctness of the hypotheses formulated above, an investigation was conducted on poly(oxymethylene) supramolecular structures obtained as a result of trioxane polymerization in different solvents and in the presence of $BF_3 \cdot O(C_2H_5)_2$ and $SnCl_4$. In all cases the thermodynamic conditions of polymerization were varied by changing the monomer concentration.

In all the solvents under investigation, the supramolecular structure of the polymer differed markedly when the process was conducted above and below the limiting equilibrium concentration of the monomer in relation to the dissolved polymer ($[M]_{lim}^{liq}$). The absolute values of $[M]_{lim}^{liq}$ vary for different solvents. At $[M] > [M]_{lim}^{liq}$ ($\mu_{M\,dis} > \mu_{P\,dis}$), the polymer consists of small (0.1 μm) irregularly shaped, often globular formations; at $[M] < [M]_{lim}^{liq}$ ($\mu_{M\,dis} < \mu_{P\,dis}$) it consists of large (5 to 10 μm and more) well-defined lamellar crystals with a shape characteristic of poly(oxymethylene) (Fig. 6). What is significant here is not the absolute value of monomer concentration, but the ratio between $\mu_{M\,dis}$ and $\mu_{P\,dis}$, i.e., [M] and $[M]_{lim}^{liq}$. Varying the concentration and nature of the catalyst and reaction time does not produce a qualitative effect on the morphology of the polymer formed. Thus, the absolute value of polymerization rate and polymer yield do not, in this case, have a decisive effect on polymer structure. Similar results were obtained through the cationic polymerization of 1,4-diazabicyclo[2.2.2]octane.

It has thus been shown that it is the thermodynamic conditions, in which the processes are conducted, that determine the supramolecular structure of the polymer

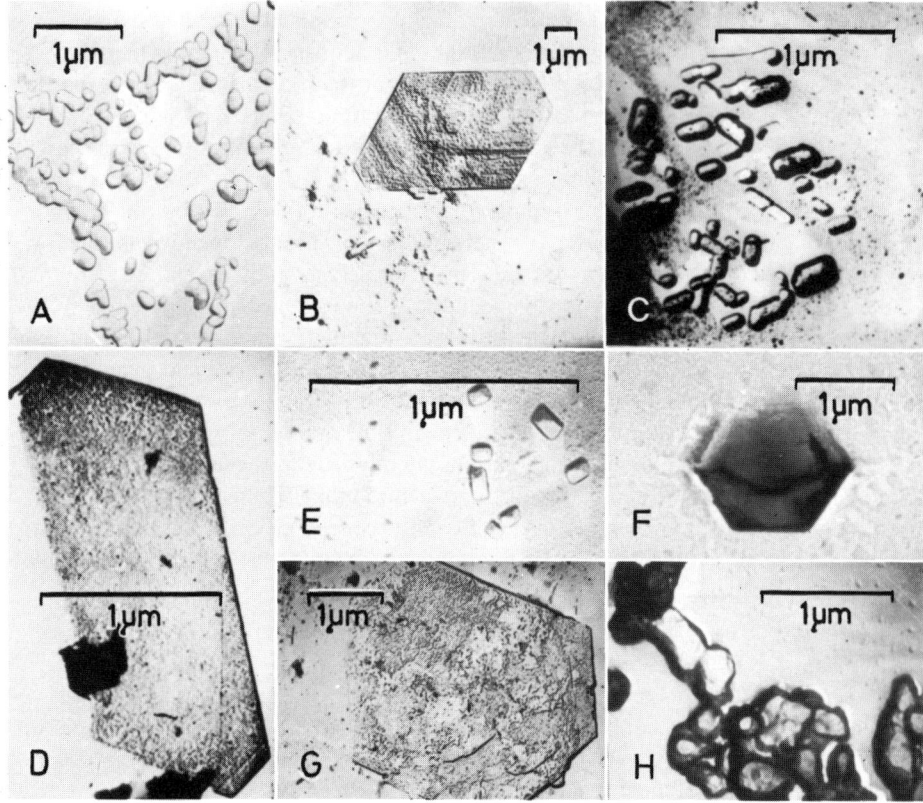

Fig. 6 A–H. Microphotographs of the polymers obtained in methylene dichloride (**A, B**); nitrobenzene (**C, D**); 1,2-dichloroethane (**E, G**) and heptane (**F, H**) at 25 °C (**A–G**) and 65 °C (**F, H**); [M] = 3 (**A, E**); 1,7 (**B**); 2 (**C, F**); 0,5 (**D, H**) and 1.0 mol/l (**G**). $c_k = 4.5 \times 10^{-2}$ (**A**); 2×10^{-2} (**B**); 4.6×10^{-3} (**C**); 5.2×10^{-3} (**D**); 3.6×10^{-2} (**E, G**); 2.8×10^{-3} (**F**) and 1×10^{-3} mol/l (**H**). c_k – catalyst's concentration in mol/l

formed. Moreover, results of investigations into poly(oxymethylene) morphology in trioxane polymerization confirming the finding of the process mechanism resulting from an analysis of kinetic data. Indeed, according to these results, at $[M] > [M]_{lim}^{liq}$ the chain propagation takes place on dissolved active centers with the formation of a polymer chain in solution and is followed by its separation into an individual phase and crystallization. The crystallization and formation of the polymer phase in this case occur during an apparent overcooling (or oversaturation), whereupon a polymer of imperfect structure is formed: at $[M] < [M]_{lim}^{liq}$ the chain propagation takes place directly on the surface of the polymer crystal. The molecule being added can attach itself to, and disengage itself from, the end of the chain many times before it finally "chooses" the most favorable site. This behaviour results in the formation of a perfect polymer monocrystal.

The same thermodynamic approach can be extended to the control of the molecular structure of polymers. With regard to molecular structure, the formation of a regular polymer, such as a syndiotactic or isotactic polymer, or a regular copolymer of the $(A_a B_b)_p$ type, is usually an entropically unfavorable process. This statement is true when we speak of isolated macromolecules, or when we compare polymers in the amorphous phase or in solution.

If, however, conditions favor the crystallization of a regular polymer, the chemical potential of this polymer can under identical conditions become lower than under the chemical potential of an irregular polymer because of the gain in crystallization energy. Indeed, if, for example, the units of a copolymer were not connected by chemical bonds, they would, given the appropriate conditions, crystallize into separate phases. Therefore, when the nature of the solvent changes, and it no longer dissolves both the regular and irregular polymers, but only the regular crystalline polymer, a radical change in the relationship between the chemical potentials of polymers of different structure can result.

Within the framework of the approach formulated above, the copolymerization of trioxane and 1,3-dioxolane in methylene dichloride and nitrobenzene was studied at different absolute concentrations of monomers and different concentration ratios. The catalysts used were $BF_3 \cdot O(C_2H_5)_2$ and $SnCl_4$. At high trioxane concentrations, i.e., in reaction conditions corresponding to $\mu_{T\,dis} > \mu_{T\,cp\,dis}$ (where $\mu_{T\,dis}$ and $\mu_{T\,cp\,dis}$ are the chemical potentials of trioxane and of the trioxane units in the dissolved copolymer), both solvents displayed a "normal" dependence of copolymer composition on the composition of the monomer mixture. Thus, with an increasing amount of dioxolane in the monomer mixture, its percentage in the copolymer increases substantially. In this case, as with trioxane homopolymerization under similar conditions ($[M] > [M]_{lim}^{liq}$), the active center seems to be in a dissolved state. Copolymer composition is, therefore, determined by the laws of homogeneous copolymerization.

At low trioxane concentrations, i.e., at $\mu_{tr.dis} < \mu_{tr.cp.dis}$, the copolymer composition does not depend on the composition of the monomer mixture. The average length of dioxolane blocks is also constant and differs markedly from the length expected in a random distribution of dioxolane units in the copolymer.

Morphological studies of copolymers have shown that, in the first case, they have a globular structure, while in the second, lamellar hexahedral copolymer monocrystals are formed. All these investigations show that, regardless of the ratio between the monomers in the mixture, only the quite definite, and most probably regular, polymer obtained is the most favorable from a thermodynamic point of view. The internal portion of the copolymer monocrystal can be expected to consist of trioxane units, with the dioxolane units located on the lower and upper surfaces of the monocrystal and forming a fold in the chain.

The proposed thermodynamic method thus makes it possible to control both the supramolecular and molecular structures of polymers. Potential uses for this method seem to be far from exhausted. In particular, it can be used in thermodynamic analyses of the supramolecular structures of polymers, and in the synthesis of stereoregular polymers and regular cross-linked polymers.

II. Polymerization Kinetics of Heterocycles

In a series of papers[43–45], the kinetics of anionic polymerization of ethylene oxide in conjunction with different catalysts were studied. These studies expand our understanding of the mechanism of living polymerization systems and provide new information on the processes of active center association. Herein, primarily, lies the specific nature of the heteroatomic systems, as compared with the vinyl monomers studied earlier[9].

The processes which were investigated can be subdivided into three groups: (a) homogeneous processes with catalysts containing a counterion based on an alkali metal; (b) homogeneous processes with catalysts containing a counterion based on an alkali-earth metal; (c) heterogeneous processes on solid catalysts.

Homogeneous polymerization was shown to proceed by a living mechanism, with the number of active centers equal to the initial catalyst concentration. The initiating systems were K-, Na-, and Cs-naphthalene.

A specific feature of these systems, as compared to vinyl monomers, is a reduced predisposition to dissociation and an enhanced predisposition to the association of ion pairs. This is explained by a greater localization of charge on the end oxygen atom of poly(ethylene oxide), as compared to the carbon atom in the polymerization of vinyl compounds.

The only type of active centers in the polymerization of ethylene oxide is the free ion pair. There are only a few free ions, and their activity only slightly exceeds that of ion pairs. For a concrete system in dimethyl sulfoxide (Na counterion) the ratio of rate constants is equal to 2.

The main forms in which the ion pairs exist are trimers (counterion K, Cs) and tetramers (Na). The kinetic study of ethylene oxide polymerization over a broad range of initiator concentrations established the existence of transition from the associated state of ion pairs to the free state when their concentration is reduced (Fig. 7).

Another feature of ethylene oxide polymerization is the participation of the polymer chain in counterion solvation. From this it follows that: (1) the reaction is autoaccelerated at the initial stages, and (2) the sensitivity of kinetic parameters to the nature of solvent is low.

The kinetic laws governing the polymerization of ethylene oxide in the presence of alcoholates of alkali-earth metals are similar to those described above. However, because of the high electric field intensity of doublecharged cations, the association of ion pairs in homogeneous systems is even stronger than for alkali metals. In addition the effective rate constants are much smaller than in the previous case.

In those instances when these same centers are fixed on the solid surface of the catalyst, their association is impeded, thereby raising to a considerable extent the catalytic activity of heterogeneous catalysts. Therefore, catalysts of the $SrCO_3$ or Ca(diphenyl)$_2$ type display a considerable activity and make it possible to produce high-molecular-weight poly(ethylene oxide).

The possibility of obtaining a high-molecular-weight product in ionic polymerization is usually restricted by the presence of hydroxyl-containing impurities. Heterogeneous catalytic systems turn out to be slightly sensitive to impurities. The reason

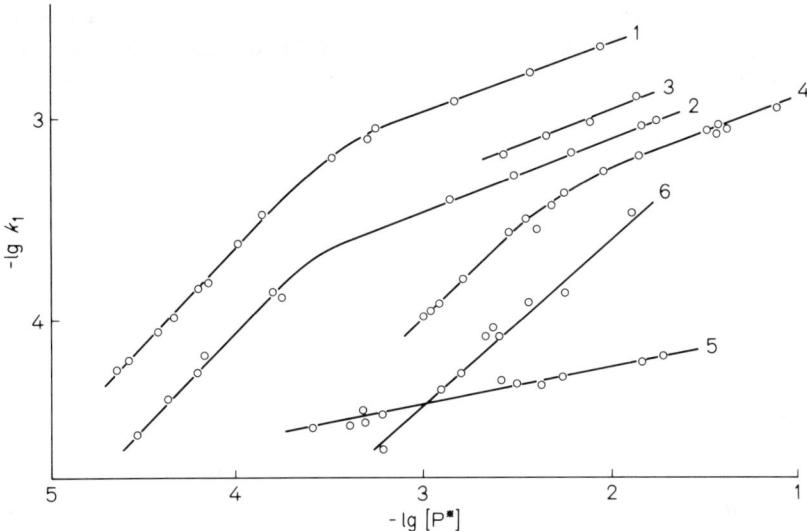

Fig. 7. Dependence of ethylene oxide polymerization rate constant on active centers concentration: 1 = Cs^+, THF, 60 °C; 2 = K^+, THF, 70 °C; 3 = K^+, DG, 70 °C; 4 = K^+, in block, 40 °C; 5 = Na^+, THF, 100 °C; 6 = K^+, DMSO, 60 °C; (unit of k_1 is s^{-1} and of [P*] mol/l), DG – 2,5,8-trioxanonane, DMSO – dimethyl sulfoxide

for this sensitivity is not yet absolutely clear, but it can be assumed that the impurities are adsorbed on the catalyst and/or interact chemically with it.

Let us now consider cationic polymerizations. Tetrahydrofuran and sulfur-containing cycles were known to undergo cationic "living" polymerization[46, 47]. It was later found[48–52] that nitrogen-containing cycles undergo cationic polymerization in methanol following the same mechanism.

When the reaction is initiated with haloid alkyls or quaternary ammonium salts of monomers (the latter imitate the active center of a polymer molecule), the initiation rate is sufficiently high ($k_i \geqslant k_p$) to obtain high molecular-weight polymers. This is indicated by the following kinetic regularities observed: the first order of the polymerization rate with respect to the initial monomer concentration and the course of the process, and approximately the first order with respect to the catalyst concentration. When a fresh portion of the monomer is added after the process is terminated, polymerization resumes at practically the same rate, and the molecular weight of the polymer increases. Molecular weight of the polymer grows in proportion to the initial monomer concentration and conversion level; the slope of the line in $P_n - q\,[M]_0$ coordinates is always equal to the inverse value of the catalyst concentration, regardless of temperature, monomer concentration, or the nature of the counterion. The measured rate constant is thus the effective constant of the chain propagation rate. The chain propagation seems to be actualized by the nucleophilic attack of the monomer on the active center, the quaternary ammonium salt of the monomer. This is indicated by the kinetic data, as well as by the fact that polymerization proceeds at high rates and without termination with such counterions as OH^-, $C_6H_5O^-$, and F^-. The polymerization of nitrogen-containing cycles through the action of acids or tertiary ammonium salts is also a continuous process. In this case, however, the finding that the initiation rate constant is by two orders of magnitude lower than the propagation rate constant is in agreement with well-known data on the alkylating capacity of quaternary and tertiary ammonium salts.

The observed dependencies of molecular mass on degree of conversion q, initial concentrations of the monomer ($[M]_0$) and the initiator ($[I]_0$), as well as the presence of an induction period in the kinetic curve, are in agreement with the reaction scheme which takes into account slow initiation:

$$\bar{P}_n \sim \sqrt{\frac{q\,[M]_0}{[I]_0}}$$

Similar kinetic results have also been obtained when the polymerization was initiated by a BF_3 complex with a monomer.

By NMR on protons, fluorine, and boron, and by conductivity studies, the initiating particle structure was unambiguously determined in this case as:

$$\begin{array}{c} CH_2-CH_2 \\ HN^+ \quad CH_2 \\ CH_2 \; CH-CH_2 \\ CH_2 \end{array} \qquad BF_3OCH_3^-$$

In all cases the chain propagation rate constants coincide for one and the same counterion and one and the same catalyst concentration, regardless of the nature of the initiating particle (acids, alkyl halides, quaternary and tertiary ammonium salts). This finding confirms the correctness of the assumptions on the mechanism and indicates that the active centers have a common structure.

Conductivity studies of quaternary ammonium salts and polymerization systems, experiments with additions of neutral salts, and investigations on the counterion effect make it possible to state that free ions and ion pairs serve as active centers in the polymerization of nitrogen-containing cycles and also to separate the contributions of the former and the latter to the effective chain propagation rate constant (see Fig. 8).

As seen in Fig. 8, free ions are less active than ion pairs, and the higher the size of the counterion, the higher the rate constant. A similar effect is produced by the size of the cation itself: the ratio between the constants of propagation rate of the ion pair and the free ion is higher for a four-membered cycle than for a three-membered one with the same counterion. The counterion seems to take an active part in the transition complex whose structure is the more favorable the larger the size of participating particles becomes.

A change in the polarity of the system (methanol/1,4-dioxane mixture) results primarily in a change in the degree of dissociation of ion pairs and, as a result, in a change in the reaction rate. Since ion pairs in these processes are more active than free ions, a decrease in the polarity of the medium results in an increase in the polymerization rate, all other things being equal.

Kinetics of Polymerization Processes

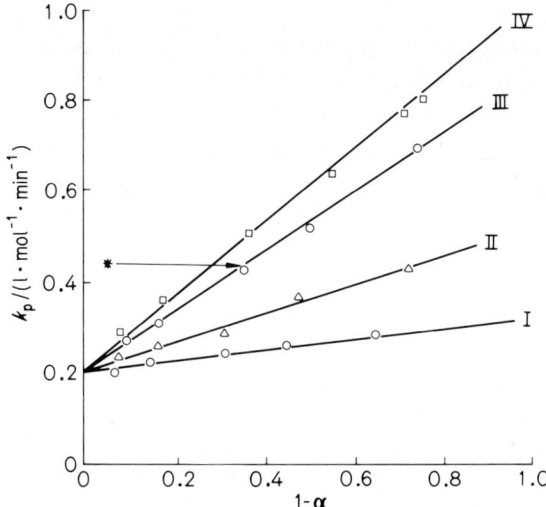

Fig. 8. Propagation rate constant of conidine polymerization vs. degree of salt dissociation. I) N-Ethylconidinium chloride; II) bromide; III) iodide, and IV) perchlorate. CsI salt (10^{-2} mol/l) was added. The arrow shows the degree of dissociation corresponding to this concentration. Monomer concentration: 6 mol/l; $T = 60\,°\text{C}$

These studies have thus made it possible to determine the kinetic parameters of polymerization on free ions. The thermal effect of the reaction was measured calorimetrically. These results are presented in Table 1:

Table 1

Monomer	k_p^+ $l \cdot mol^{-1} \cdot min^{-1}$	E kcal/mol	ΔS cal/(mol · K)	$\Delta H°_{liq\text{-}liq}$ kcal/mol
CH₂–CH₂–N(–CH₂–)CH–CH₂ (conidine)	$k_{40\,°C} = 2.7$	9.1	−38	−23.4
bicyclic N(CH₂–CH₂)₃CH–CH₂	$k_{60\,°C} = 0.2$	11.4	−38	−16.8
N–CH₂–CH₂–N(CH₂–CH₂)(CH₂–CH₂)	$k_{195\,°C} = 0.125$	19.2	−35	−2.3

In papers[53, 54] an attempt has been made to investigate the mechanism of cationic cyclic acetal polymerization of 1,3-dioxolane. The presence of an acetal bond in the monomer molecule decreases the stability of active centers which are subjected to various reversible and irreversible chemical transformations.

The concentration of the functioning active centers for example turns out to be much lower than the initial catalyst concentration. When initiation takes place by Lewis acids and oxonium salts, the concentration of active centers is determined by the "catalyst − active centers" equilibrium, which is established during a period of time much shorter than the overall time of polymerization.

A number of experimentally observed facts are direct consequences of the ascertained mechanism: (a) it has been established that the addition of substances displacing the equilibrium toward the formation of active centers produces a cocatalytic effect; (b) cyclic acetals are polymerized following the mechanism of "living" polymerization, but at each given moment only a small percentage of the total number of macrochains propagate. The other chains are "enlivened" in the course of polymerization. The number of macromolecules is determined by the equilibrium substitution of ligands in Lewis acids and by other equilibrium reactions, which explains the well-known linear increase of the molecular mass of acetals with an increasing degree of conversion. (c) The absence of an unambiguous effect of dielectric constant, ϵ, on the rate of the process, which is characteristic of the polymerization of cyclic acetals, is a direct consequence of the competition between two factors: a decrease in k_p with growing ϵ and an increase in the equilibrium concentration of active centers.

Researches previously attempted to treat the mechanism of the polymerization of cyclic acetals as being similar to the mechanism of tetrahydrofuran (THF) polymerization. The above data show an essential difference between the cationic polymerization mechanisms of cyclic ethers and cyclic acetals.

In the joint polymerization of 1,3-dioxolane and THF, a loss of catalytic activity with time is observed with respect to THF. This activity is preserved, however, with respect to dioxolane and may be a consequence of counterion decomposition in the acetal medium, along with a higher reactivity of acetal as compared to ether.

In papers[55,56] the authors studied the effect of water on the polymerization of cyclic acetals. It was concluded that the inhibiting action of water during the induction period is caused by the direct chemical interaction of water with the active centers.

Several papers[57−59] were devoted to investigating a complex process such as the cationic copolymerization of monomeric formaldehyde with dioxane in the gas, liquid, and gas-liquid phases. It is known that polyacetal resins are industrially produced by copolymerizing cyclic acetals (trioxane, 1,3,5,7-tetraoxane), or by anionic homopolymerization of monomeric formaldehyde with subsequent modification of end groups.

The cationic polymerization of monomeric formaldehyde with cyclic acetals and ethers, which bypasses the stage of cyclic formaldehyde oligomer formation, has been known some time[60, 61]. It was, considered difficult, however, to introduce a process of this kind into industry because of the high sensitivity of the reaction system to polar impurities.

Nevertheless, an industrial process of this type[62] has been instituted in the USSR. The structure and properties of the manufactured product are somewhat different from those conventionally obtained: they lie between those of trioxane copolymers of the "Hostaform C" type and acetylated homopolymer of the "Delrin" type.

New investigations into the kinetics of monomeric formaldehyde copolymerization were subsequently conducted. In papers[63, 64] a degradation kinetics method for the analysis of polyacetal microstructure is proposed; in paper[65] is outlined a pyrolitic method based on a chromatographic determination of 1,3,5-trioxepane at the junctions of oxymethylene and dioxolane blocks.

Independent determinations of microstructure and composition heterogeneity of polyacetals, which were obtained using various methods and based on different monomers (formaldehyde and trioxane to produce oxymethylene blocks, and methylene oxide and dioxolane to produce stabilized oxymethylene blocks), made it possible to reveal differences in the mechanism of the synthesis of such products.

The trioxane copolymer, for example, differed from the formaldehyde copolymer by a narrower distribution of C–C bonds (80.3% of individual blocks against 45–50%) and a more uniform distribution of these bonds among the copolymer fractions. At the same time, the high-molecular-weight fractions of the formaldehyde copolymer had very few C–C bonds; this is what caused the increased thermal stability and melting temperature of this product, as compared with the trioxane copolymer, which has the same overall content of C–C bonds.

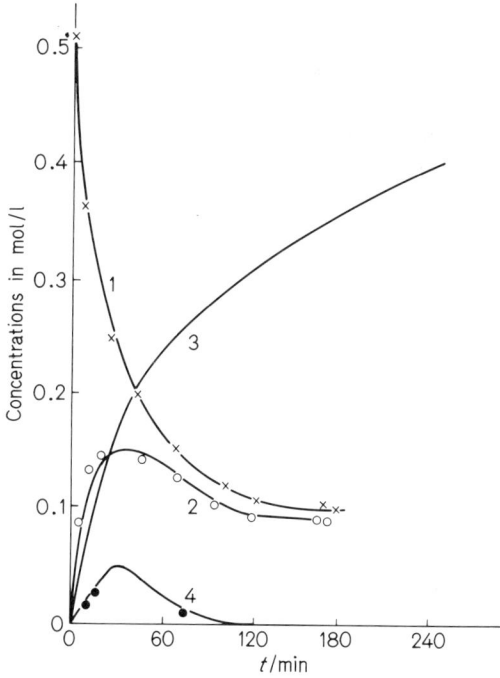

Fig. 9. Kinetic curves of the reagents conversion in liquid-phase cationic copolymerization of formaldehyde with 1,3-dioxolane. 1) 1,3 dioxolane consumption; 2) 1,3,5-trioxepane yield; 3) yield of ethylene oxide units in copolymer; 4) yield of soluble polymer

The process of formaldehyde copolymerization with dioxolane is characterized by the formation of a number of by-products with a dynamic equilibrium established among them[60]. To the previously found trioxepane and trioxane was later added a soluble copolymer whose composition is similar to that of polytrioxepane[58] (Fig. 9).

In a series of papers[58, 66, 67], it was established that this process represents a special type of copolymerization in which a statistic copolymer is formed in spite of the fact that formaldehyde and dioxolane are polymerized on different types of active centers.

III. Macrozwitterionic Polymerization

Assumptions concerning a possible zwitterionic mechanism were made by a number of authors for different cases of anionic and cationic polymerization[68–73]. This mechanism was well substantiated for the anionic polymerization of diethyl vinylmalonate under the action of triphenylphosphine[68], and β-propiolactone under the action of trimethylamine[69] and betaine[70].

A similar mechanism was also proposed for the cationic polymerization of a number of heterocycles and aldehydes under the action of Lewis acids, but the researchers reasoning was less convincing.

In papers[71–73] the kinetics of acrylonitrile and methacrylonitrile polymerization on triethylphosphine was studied in detail.

Both monomers are polymerized in accordance with the ionic mechanism. Elemental analysis showed the phosphorus to be chemically bonded to the polymer chain, with one initiator molecule per each polymer chain.

By NMR on ^{31}P and ^{1}H nuclei, it was shown that phosphorus is present in the polymer in the form of a quaternary phosphonium group, the zwitterion structure was also established. With time, the electric conductivity of the system increases in the presence of water, which is indicative of slow initiation. In the absence of hydroxyl-containing compounds, the spontaneous annihilation of active centers is possible. The steady-state concentration of active centers is low, and it is thus possible to disregard the interaction of zwitterions with one another.

The equations for the initial reaction rate and $\bar{P}_w$, proceeding from the kinetic scheme proposed

$Et_3P + M \longrightarrow Et_3P^{\pm}-CH_2-\bar{C}HCN$ (initiation)
$Et_3P^+-CH_2-\bar{C}HCN + nM \longrightarrow Et_3P^+-(CH_2-CHCN)_n-CH_2-\bar{C}HCN$ (propagation)
$Et_3P^+-(CH_2-CHCN)_n-CH_2-\bar{C}HCN + R'OH \longrightarrow$
$R'O^-[Et_3P^+-(CH_2-CHCN)_n-CH_2-CH_2CN]$ (annihilation)

have the following form:

$W_0 = k_i k_p [C]_0 [M]_0^2 / k_t'$, $\bar{P}_w = 2(k_p/k_t')(1-q)$

where k_t' is the constant of chain termination rate on the hydroxyl (of the pseudo-first order).

To explain the observed fourth order of the reaction in relation to the monomer, the authors assumed it was caused by the permittivity of the medium, since the monomer is more polar than tetrahydrofuran, the solvent.

Kinetics of Polymerization Processes

Adding a high-molecular-weight solvent (dimethylformamide) indeed raised the rate and the polymer molecular weight. On the other hand, when the reaction was conducted under isodielectric conditions, the second order of reaction was approached.

An isolated polymer zwitterion can exist in solution in two conformations:

ring (cycle)   coil

The distance between the ends of the convoluted macromolecule in solution is described by a Gaussian normal distribution; the longer the polymer chain, the greater the distance between its ends. Electrostatic interaction between the ends of the zwitterion makes the existence of an "ion pair" hypothetically possible.

The probability of the existence of active centers in the form of ion pairs or free ions is determined by the free energy variation due to (a) a change in the entropy and transition from Gaussian distribution to "cyclic" conformation and (b) a change in the energy of Coulomb interaction at this transition. This probability evidently depends on the length of the macromolecular chain.

The constant of the equilibrium of free ion – ion pair transition can be written thus

$$\ln K = \Delta F_e - \Delta F_c/(kT)$$

Here
ΔF_e is the change in the free energy of ions and ion pairs at the changeover of the active centers;
ΔF_c is the change in the polymer chain free energy at this changeover.

According to the statistics of polymer molecules,

$$\Delta F_c = kT \ln \left[\left(\frac{3}{2 \pi n l^2} \right)^{3/2} V_s \right]$$

where V_s is the effective volume of ion pair, l and n are the length and number of segments in the polymer zwitterion.

From these equations it follows that the equilibrium constant of polymer zwitterion at the ion pair – free ion monomolecular transition is equal to:

$$K_p = K \cdot n^{3/2}$$

The probability of encountering an active center in the form of a free ion increases with increasing length of the polymer zwitterion.

For vinyl monomers the constant of propagation rate on free ions is higher than on ion pairs. Therefore, with increasing conversion in polyacrylonitrile (PAN) polymerization, the polymerization rate must increase.

PAN is, however, only slightly soluble in the reaction medium. Therefore, to verify the hypothesis, a homogeneous system was used: methacrylonitrile/triethylphosphine/dimethylformamide, which is polymerized in accordance with the same zwitterionic mechanism. The "dead" macromolecule preserves a stable positive charge. This means that the possibility for the growing free anion to interact with a "foreign" counterion must be taken into account. This interaction will grow in relation to the accumulation of polymer in the system.

Among the experimental findings were the inhibiting effect of the polymer, the yield restriction in the course of the reaction, and the decelerating effect of salts dissociating during dissolution. These results can be explained by the displacement of equilibrium toward the formation of "mixed" ion pairs.

The kinetic scheme of the process was examined in a steady-state approximation under the condition that at the beginning of the process the predominant role belongs to the monomolecular equilibrium of active centers, meaning that an ion pair is formed only with "its own" counterion.

The fraction of free ions is then equal to:

$$[R_n] = K_p^{(n)} [r_n]$$

where $[r_n]$ is the concentration of ion pairs.

The formal order of the initial polymerization rate in relation to the monomer turned out to be equal to 2.5, and the molecular mass was proportional to $[M]_0^{1.7}$.

The molecular mass of the product reaches its maximum at $q \approx 0.15$ and then decreases (see Fig. 10). This decrease is caused by the influence of the polymer which has accumulated in the system and by the corresponding shift of equilibrium toward inactive ions.

Taking into account the dependence of free ions/ion pairs equilibrium on the chain length, analysis of the nonsteady-state portions of kinetic curves, yields the following equation for the induction period:

$$\tau = \frac{1}{k_r} + \frac{2}{k_{(-)} \cdot K_p^{(1)}} \cdot \frac{1}{[M]_0}$$

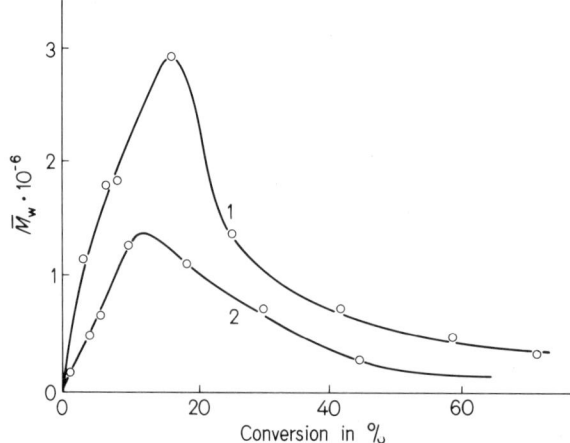

Fig. 10. Dependence of weight average molecular mass of poly(methacrylonitrile) on conversion. Monomer concentration: 1 = 1,2; 2 = 2,1 mol/l; [Et$_3$P] = 10 mmol/l

where $k_{(-)}$ is the constant of propagation rate on free ions;
$K_p^{(1)}$ is the constant of free ions/cyclic ion pairs equilibrium.

From this equation it follows that the induction period is inversely proportional to $[M]_0$ and does not depend on the initiator concentration. This is confirmed by the data in Fig. 11.

From the dependencies obtained, as well as from those previously obtained for the steady-state regions of the relationships of elementary constants, the following values have been found:

k_i = 5.6 × 10^{-4} l · mol^{-1} · min^{-1}

$k_{(-)}$ = 2.5 × 10^4 l · mol^{-1} · min^{-1}

$k_{(\pm)} \leqslant 2.0$ l · mol^{-1} · min^{-1} (propagation rate constant for cyclic ion pairs)

The reactivity of free ions proved to be much higher than that of ion pairs $(k_{(-)} \gg k_{(\pm)})$.

The dependence of the effective propagation rate constant on the polymer chain length is expressed by the equation

$$k_{eff} = (k_{(\pm)} + k_{(-)}K_p^{(n)})/(1 + K_p^{(n)})$$

This dependence is shown in Fig. 12.

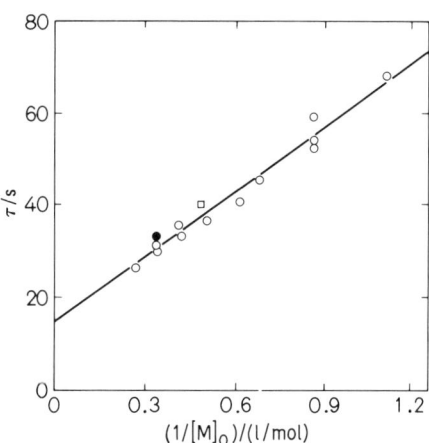

Fig. 11. Dependence of the polymerization induction period τ on reciprocal initial monomer concentration $1/[M]_0$

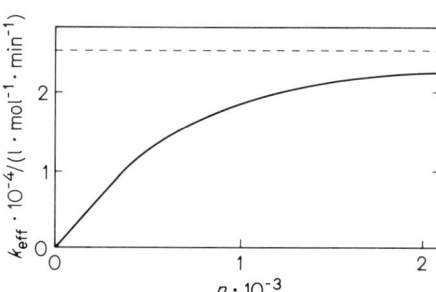

Fig. 12. Dependence of the effective propagation rate constant on polymer chain length n

With the growth of the polymer zwitterion, the rate of monomer addition is to a growing extent determined by the chain propagation on free ions.

C. Modeling the Kinetics of Polymerization Processes

Recent progress in computer technology and methods for rapid experimental determination of molecular weight distributions (MWD) has encouraged the development of mathematical simulations of polymer synthesis processes in various types of batch and continuous reactors.

Apart from the purely practical importance of such calculations, these advances have proven useful in solving various kinetic problems and considerably expanded the possibilities for further investigations, primarily because MWD is now included among the kinetic parameters of polymerization.

At the same time, the inordinate fascination with the methods of finding the values of kinetic constants by solving inverse problems can also lead away from a true understanding of the deeply rooted fundamentals, even though they formally provide a means of "filling in the gap" in investigation methods.

Several recently published reviews[74–76] sum up the results of investigations in this field, but the individual works differ in value. Three problem areas concerning modeling are examined below. These include methods for determining the kinetic constants, analysis of nonisothermal processes, and applied calculations.

I. Application of the Mathematical Simulation Methods to Verify the Kinetik Scheme and the Values of Kinetic Constants

Comprehensive calculations of radical polymerization processes, including the high conversion degrees of monomers, necessitate taking into account the dependence of the values of kinetic constants on the viscosity of the medium. However, neither theoretical calculations in relation to the proposed models (see Section A), nor attempts at a direct experimental measurement of the values of constants as a function of viscosity yield sufficiently reliable data for applied calculations of polymerization processes.

In one of the earlier works on the modeling of the radical polymerization accompanied by autoacceleration[77], the investigators tried to take into account the gel effect by introducing the empirical function

$$(k_t)_q = (k_t)_0 \, [M]^Z$$

where $Z > 0$ is a constant.

In another work[78] a semiempirical equation in the following form is proposed:

$$k_t \, (mn) = BD_1 \, (m^{-1/2} + n^{-1/2})$$

where B is a coefficient, D_1 is the diffusion constant of the monomer, and m and n are the degrees of polymerization of molecules undergoing recombination.

In both cases, however, the data obtained are purely qualitative.

A better agreement between the theoretical and experimental data was obtained in[79], where styrene oligomerization in carbon tetrachloride was studied. Five different variants of the functions interpolating the dependence of the chain propagation constant on chain length were computed, with k_p regarded as a constant. Therefore, the relations were only empirical.

Thiele et al.[80] modeled the initial polymerization of styrene. A comparison with the experiment showed that there is a satisfactory correlation of W and $\bar{P}_n$ up to $q \leqslant 0.4$. On the basis of the literature and experimental data, it has been concluded that k_t in isothermal suspension polymerization becomes diffusion-controlled at $q > 0.25$; initiation efficiency (f) becomes diffusion-controlled at $q > 0.4$; and k_p and k_m become diffusion-controlled at $q > 0.75$.

Accounting for diffusion hindrances in styrene suspension polymerization in an industrial reactor showed only an indirect correlation with respect to MWD obtained by the fractionation method.

The empirical dependencies of k_t and f on the viscosity of the medium proposed by Hamielec et al.[81-83] made it possible to model the process of styrene initiated polymerization up to $q = 0.6$.

Verification of the mechanism of styrene thermo-initiated polymerization in the initial stage of the process is reported in[84]. Assuming the presence of a chain transfer to one of the adducts formed during initiation, the authors attempted to find an explanation for the anomalous growth of molecular weight at the very beginning of polymerization.

For some values of the constant of chain transfer to the adduct ($k_{ad} > k_m$), a rather good agreement between the calculated and experimental $\bar{P}_n = f(q)$ relationships was observed, but only up to a conversion of $q \leqslant 0.08$. If, however, one tries to extend this mechanism to higher conversions, $\bar{P}_n$ according to the model increases, whereas in experiments it does not depend on conversion.

In the experiments of Hamielec et al.[85] and of Soviet researchers[86, 87], the method of solving an inverse kinetic problem has been used to find the dependence of values of effective constants on conversion in the thermo-initiated bulk polymerization of styrene.

In both cases the researchers started from the initiation mechanism proposed by Mayo[84]. Hamielec et al. examined the second and third orders with regard to the monomer during initiation and obtained the dependence of kinetic constants on conversion for the isothermal process in the following equations:

$$\frac{W}{[M]} = \left(\frac{W_i k_p^2}{k_t}\right)^{1/2} = A_0 \exp(A_1 q + A_2 q^2 + A_3 q^3)$$

$$C_m = \frac{k_m}{k_p} = (C_m)_0 + B_1 q$$

The initial values of W_i and kinetic constants were taken from the literature; the values of coefficients A_0, A_1, A_2, A_3, and B_1 were found in the experimental data using a computer $(q - t, \overline{P}_w/\overline{P}_n = f(q))$.

Isothermal bulk polymerization of styrene is satisfactorily described by the model up to the limiting concentrations in the 100–200 °C temperature range. The model was later extended to cover the 200–230 °C temperature range without additional accounting for the reactions of polymer degradation[88].

The Soviet investigators undertook the task of modeling the process of nonisothermal styrene polymerization in the same temperature range. Starting from Mayo's kinetic scheme, they reduced the polymerization mechanism to four constants (k_i, k_p, k_t, k_m). The availability of two independent experimental dependencies $(W = f(q))$ and $\overline{P}_n = \varphi(q))$ makes it possible to determine only two values of constants for each quantity.

Kinetic analysis has shown that the experimental data can only be described if $k_p = f(q)$, $k_t = f'(q)$, $k_i = $ const, and $k_m = $ const. Although it is difficult to explain the finding of asymbathic k_p and k_m variations within the framework of the physical interactions between chain propagation and transfer to the monomer, the researchers obtained a satisfactory agreement between the calculated and experimentally measured MWD at P_w/P_n varying from 1.92 to 6.0.

Initial values of kinetic constants were taken from the literature. The model unsatisfactorily described the experiment at $T \leqslant 100$ °C.
The equations for k_t and k_p are as follows:

$$k_t(q) = \exp(-1.433\, q^2 - 7.799\, q + 2.57) \exp\left(\frac{280}{RT}\right)$$

$$k_p(q) = \left(7.757 \cdot 10^{10} + \frac{8.902 \cdot 10^9}{q + 1.41}\right) \exp\left(-\frac{6800}{RT}\right)$$

The nature of the $k_p(q)$ dependence was determined by the initial value of W_i. When this value changed by an order of magnitude, the dependence of k_p on q disappeared.

Since the experimental determination of the correctness of the model included the measurement of very broad bimodal MWD, the researchers used three parallel methods: gel chromatography, chromatographic fractionation, and the method of thermal precipitation with turbidity recorded by light scattering.

Characteristic kinetic and morphological features of vinyl chloride radical polymerization processes were reviewed in[89]. While developing a mathematical model for the polymerization of this monomer, Canadian[90] and Soviet[91] investigators concentrated their attention on the heterophase nature of the process. In both cases the dependence of kinetic constants on the viscosity of the medium was disregarded.

When, polymerization kinetics was analyzed at high degrees of conversion, however, there was a discrepancy between the calculated and experimental kinetic curves.

Kinetics of Polymerization Processes

As a result in[92] an attempt was made to verify the values of kinetic constants during the final stages of the process. A special installation was designed to investigate vinyl chloride adsorption by the polymer from the gaseous phase.

The limiting concentration of the monomer in the polymer phase was found to be equal to 23%. This figure is in good agreement with the data on the ultimate swelling of the polymer recorded in the literature.

The polymerization kinetics in the polymer phase at its constant composition and continuous monomer supply was described by the equation

$$\frac{d[P]}{dt} = k_p [M] \sqrt{\frac{2 k_2 \cdot f_2}{k_t} \cdot [I]_2^0 \left(V_p + \frac{1-\alpha}{\alpha} V_m \right) e^{-k_2 t}}$$

where

[P] is the polymer concentration; k_p, k_t, and k_2 are the constants of chain propagation, chain termination, and initiation in the polymer phase, respectively; f_2 is the initiation efficiency; [M] and $[I]_2^0$ are the concentrations of monomer and initiator (initial) in the polymer phase; V_p and V_m are the molar volumes of the polymer and the monomer; α is the weight fraction of polymer in the polymer phase.

From this equation the values of k_2 were found by plotting a semilogarithmic anamorphosis, $2.3 \lg \left(\frac{dx}{dt} \cdot \frac{1}{\sqrt{x}} \right) - t$, for given temperatures and polymer phase compositions.

With a decreasing content of monomer in the polymer phase, k_2 proved to decrease while F grew. This agrees with the known data on the influence of the viscosity of the medium on the kinetics of initiator decay.

Canadian researchers published two papers on the modeling of methyl methacrylate radical polymerization, which is characterized by a pronounced gel effect.

Hamielec et al.[93] continue to develop the approach based on calculating the kinetic parameters from experimental data obtained by the GPC method. As with styrene polymerization, the dependence of the values of kinetic constants on conversion is sought as a complex function

$$a_1 = \frac{k_t}{k_p^2} \cdot \frac{1}{[M]_0} \cdot \frac{dq}{dt}, \text{ where } a_1 = \exp(A + Bq + Cq^2 + Dq^3)$$

found by computer from the data on instantaneous MWD and the reaction rate.

The investigators single out two stages in the process of isothermal polymerization. Diffusion control of elementary stages is postulated for the second stage of the process.

The possibilities of formally and adequately describing a set of experimental data obtained under certain conditions have been demonstrated in[94-96]. Proceeding from the same experimental data as in[94], the researchers propose another kinetic model. It is based on the assumption that, after the gel point, two types of radicals are present in the system: "entangled" ones with a limited mobility, and short ones unaffected by diffusive effects.

It is also assumed that k_p for both types of radicals are equal while k_t are different, including the cross-termination stage. The value of k_{te} for "entangled" radicals is shown inversely proportional to $\bar{P}_n$ and the volume fraction of the polymer, Φ_p, and k_{tp} of cross-termination is expressed as

$$k_{tp} = (k_t \cdot k_{te})^{1/2}$$

By solving the inverse kinetic problem for the conversion vs. time dependence, polymer concentration, [P], at the point of transition from the first to the second stage, and β_0 are determined in

$$(k_{te}/k_t)^{1/2} = \beta_0 \left(\frac{[P]}{\Phi_p \cdot \bar{P}_n^{0.5}} \right)^{1/2}$$

With the help of this model the researchers obtained a satisfactory agreement with experiments studying the polymerization of methyl methacrylate in bulk and ethyl methacrylate in bulk and in dimethylformamide solution.

A complicated problem is the modeling of branched polymerization. Theoretical aspects of this problem, including the choice of the most suitable mathematical methods and the determination of branching parameters, can be found in the studies of Bamford and Tompa[97], Zaidel and Katz[98], Zeman and Amundson[99], and Graisley et al.[100].

An important application of this approach is the radical polymerization of ethylene at high pressure. In[101] an attempt is made to model a process which takes place in an industrial mixing reactor. It appears that by solving an inverse kinetic problem, the researchers have verified the pre-exponential factors and the E values for a series of elementary stages, although what these stages are is not mentioned in the paper. This enabled the investigators to compute the dependence of $\bar{P}_n$ on conversion, temperature, concentration of reagents, and pressure.

The kinetic scheme of this process was studied in greater detail by German researchers[102]: the formation of long-chain branches (chain transfer on the polymer) and short-chain branches (intramolecular chain transfer) was taken into account.

Nine elementary constants of the model were verified by solving an inverse kinetic problem. The values obtained are not cited. The investigators note that the calculated MWD values agree, though poorly, with those obtained experimentally by the GPC method.

In[103] the author restricted himself to determining eight values of kinetic constants using the method of solving an inverse kinetic problem. Again the data are not cited. The model makes it possible to calculate the number of vinylidene groups and the long- and short-chain branches per 1000 carbon atoms. Divergence from the experimental data reaches 20%.

In the study[104], the homogeneous region of a high-pressure nonisothermal tubular reactor operation was examined.

All the publications in this field are characterized by a secrecy with regard to the values of kinetic constants, which are probably regarded as confidential information of know-how.

Another typical process involving branched radical polymerization is the production of poly(vinyl acetate). In the experiments of Stein[105, 106], the method of mathematical simulation has been used to evaluate the effect of longchain branches on the width of MWD. The reactions of chain transfer to the polymer and polymerization by the terminal double bonds of the polymer were examined separately. A comparison of the calculated and experimental $\bar{P}_w/\bar{P}_n - f(q)$ dependencies yielded the values of $C_p = k_f/k_p$ and $K = k'_p/k_p$.

For the chosen value of $C_p = 1.8 \times 10^{-4}$, a certain systematic deviation was observed in direct calculations, which the authors attribute either to the dependence of the values of constants on viscosity, or to the incorrectness of the assumption that the probability of chain transfer to the polymer is the same for all the units in the chain.

In[107] the value of $C_p = 1.2 \times 10^{-4}$ was specified on the basis of the above model. In[108] another variant of the model was computed, in which it was assumed that the terminal double bonds of the polymer have a differing activity. In study[109] the C_p value is shown to vary within a range of $0.6 \times 10^{-4} - 2.36 \times 10^{-4}$ with the changing ratio monomer : solvent.

An important conclusion from this series of investigations is that the MWD of a branched polymer in a reactor with ideal mixing will be broader than in a batch reactor. The same is true for the degree of branching.

In the Soviet study[110], the following elementary stages were taken into account in the kinetic scheme of vinyl acetate polymerization: chain transfer to the monomer, solvent, and polymer, and chain termination caused by the disproportionation of radicals. It was assumed that long-chain branches could be formed by chain transfer both to the acetate group hydrogen atoms and to the main chain hydrogen.

The dependence of termination and chain propagation constants on the viscosity η of the reaction medium was considered using the empirical equations:

$$k_t(\eta) = k_t^0 / \{0.76 + 0.24 [(\eta/\eta_p)^{0.2} + 10^{-8} (\eta/\eta_p)^{1.5}]\}$$
$$k_p(\eta) = k_p^0 / [1 + 0.65 \times 10^{-6} (\eta/\eta_p)]$$

The numerical function of the MWD and the long-chain branching (average number of branching nodes) were calculated. By comparing the calculated and experimental data on the conversion-time dependence, k_p and k_t were determined (initial k_p^0 and k_t^0 values were taken from[111]). The values of chain propagation constants were found in a similar way using MWD data. The values of constants are not cited in the work.

It is claimed that in the analysis of vinyl acetate batch polymerization in solution, a satisfactory agreement of calculations with the experiment has been obtained. In the course of polymerization, k_t became several times smaller and k_p started to decrease at $q > 0.6$.

A specific case of branched radical polymerization is the graft polymerization of styrene on rubber. This reaction is the basis for the manufacture of some industrially important composite plastics, such as shock-proof polystyrene, ABC plastic, and similar materials.

A characteristic feature of these processes is their heterophase nature, associated with the insolubility of the two polymers in a common solvent. The presence of two or more phases results in homo- and copolymerization taking place simultaneously. The problem is made more complicated by the inversion of phases taking place in the system in the course of the process and by the cross-linking of rubber macromolecules occurring at elevated temperatures.

A number of studies have attempted to model this process stage by stage and to determine the values of some kinetic constants. Thus, in[112] the researchers investigated the initial stage of isothermal bulk copolymerization of styrene with polybutadiene in the presence of di-*tert*-butyl peroxide.

The kinetic scheme includes 11 kinetic constants. From the experimental data 4 ratios of constants were determined, while the others were obtained from the literature data.

The experimental data on the $\bar{P}_n$ of homopolystyrene, the efficiency of polystyrene grafting to rubber, and the dependence of styrene polymerization on rubber concentration were compared with the calculated data, with up to 8% of styrene conversion at $T = 100\ °C$.

In the study[113] an attempt was made to model the whole process of thermal bulk copolymerization of styrene with polybutadiene up to high degrees of conversion. The calculations were based on the previously developed model of thermal bulk polymerization of styrene and supplemented with the reactions of chain transfer to rubber.

Styrene radicals were assumed to form in both phases, although until the stage of phase conversion, they were proposed to diffuse into the polystyrene phase. The model made it possible to describe satisfactorily the homopolystyrene molecular weight and MWD variations with the growing degree of conversion, up to the beginning of crosslinking between the dispersed rubber particles, which was accompanied in a number of cases by anomalous changes in the molecular weight of homopolystyrene.

It is very difficult to analyze experimentally the degree of polystyrene chain grafting to rubber because of the difficulty of separating the grafted copolymer from unreacted rubber and polystyrene occluded by rubber particles.

In the field of ionic polymerization, we should mention the investigators[87, 114] who calculated the relative chain transfer constants for the polymerization of styrene in benzene solutions and its mixture with 1,2-dichloroethane on Friedel-Crafts catalysts.

Taking into account the monomer consumption, the $\bar{P}_n - q$ curves were obtained by integrating the instantaneous values of $\bar{P}_n$.

In[115, 116] the cationic polymerization of cyclic ethers was examined theoretically and experimentally with regard to the nature of MWD variation. A theoretical analysis was made of how MWD is affected by the depolymerization reactions, monomolecular deactivation of active centers, recombination of active centers, chain transfer by hydroxyl-containing compounds, chain transfer to the monomer, and ether oxygen of the polymer chain, as well as via the end hydroxyl group.

The data obtained were used to analyze a specific system: copolymerization of tetrahydrofuran (THF) with oxiranes in the presence of BF_3. It is known that additions of water or diols to this system narrow down the $\bar{P}_w/\bar{P}_n$ of the end product from 2.0 to 1.0.

As a result of the mathematical simulation of this process, the researchers arrived at the conclusion that in this system, chain transfer reactions take place simultaneously on water and on diols which form in the course of the reaction.

In this kinetic scheme, however, the calculated values of $\bar{M}_w/\bar{M}_n \geqslant 1.2$, whereas due to a great excess of water with respect to the catalyst, $\bar{M}_w/\bar{M}_n = 1.025$. This discrepancy can be obviated if the rate constant of chain transfer on diols is assumed to drop with an increase in the degree of polymerization.

When MWD is used as an instrument for analyzing a kinetic scheme, very stringent requirements must be imposed upon the accuracy of its measurement. In[117] a method is proposed to determine the instrumental broadening function in GPC based on the obtained relationship between the corrected gel chromatograms of the initial polymer sample and its i-th fraction collected within the range of elution volumes (a_{i-1}, a_i):

$$W_0(y) = W_i(y) \int_{a_{i-1}}^{a_i} G(V, y) dV, \quad i = 1, 2 \ldots N$$

where $W_0(y)$ and $W_i(y)$ are the corrected gel chromatograms of the initial sample and its i-th fraction; G (v, y) is the instrumental broadening function. This equation, together with Tung's equation

$$F_i(v) = \int_{-\infty}^{+\infty} W_i(y) G(v, y) dy \quad i = 0, 1, 2, \ldots N$$

where $F_i(v)$ is the actual gel chromatogram, can be used to determine the instrumental broadening function. To solve the integral equation in the form of a corrected gel chromatogram, the iteration scheme of Smith-Hamielec was used. The Gaussian function of instrumental broadening and the asymmetric function

$$G(V, y) = q\sqrt{\frac{h}{\pi}} \int_y^\infty e^{-q(x-y)-h(x-v)^2} dx$$

were analyzed.

The method of reaction termination by a radioactive reagent was used to analyze the MWD of living polymer chains. By combining GPC with a scintillation counter (additional detector), one can in this case simultaneously obtain separate gel chromatograms of the overall product and of the growing macromolecules. In the latter case the numerical function of MWD is obtained.

At the same time, by separating the excess of radioactive reagents during fractionation, it becomes possible to get very accurate information on the number of active centers (up to 10^{-7} mol/l).

The application of this method to the study of THF polymerization on BF_3 + propylene oxide (bulk polymerization, $T = 20\,°C$) has made it possible to establish the following facts: in this system the active centers are deactivated in accordance with the first order reaction ($k'_d = 1.9 \times 10^{-4}\ s^{-1}$). Propylene oxide is consumed in the initiation stage according to the first order reaction. Its excess is consumed to form low-molecular-weight products.

The chain propagation rate constant, determined from the known monomer consumption values and the concentration of active centers, is $(2.8 \pm 0.4) \times 10^{-2}$ l·mol^{-1}·s^{-1}.

The data obtained for the broadening of the product MWD in the course of the reaction, combined with the theoretical calculations of the effect of various elementary reactions on MWD, made it possible to conclude that the reaction taking place in the system is that of transfer with polymer chain termination. The rate constant of this reaction is $(8 \pm 3) \times 10^{-5}$ l·mol^{-1}·s^{-1}.

In[118] carried out by Soviet investigators, the method of mathematical simulation was used to estimate the elementary constants of butadiene polymerization under the conditions of the termination of living chains on impurities.

Polymerization in solution on the butyllithium type of catalyst has been studied. In commercial butadiene the presence of allenes slowly interacting with active centers is possible. By solving an inverse kinetic problem from a comparison of experimental and theoretical conversion – time curves, the values of the constants of chain termination on the allene impurity were found. The k_f values are not cited, but it was noted that the activation energy value depends on the type of solvent.

Several studies were devoted to the modeling of coordination polymerization. In[119, 120] the researchers investigated the process of ethylene polymerization in solution under the action of the catalytic system [(C$_5$H$_5$)$_2$TiCl$_2$/(CH$_3$)$_2$AlCl]. The propagating chains were assumed to be deactivated by the disproportionation of two active centers. Monomer concentration in the reaction zone was assumed to be constant, with the kinetic constants independent of the chain length.

The discrepancy between the $\overline{M}_n$ values calculated from the model and those found experimentally reached 30%. The course of the calculated and experimental $\overline{M}_w/\overline{M}_n$ vs. t curves coincided. At first here is a rather sharp narrowing of MWD (from $\overline{M}_w/\overline{M}_n$ of 4–5 down to 2–3), and then it broadens again to 4–5 at constant polymerization temperature. The experimental curves, however, lie above those calculated.

American and English investigators in studies[121, 122] proposed a model of heterogeneous ethylene polymerization on Ziegler-Natta catalysts, which takes into account the dependence of chain termination rate constants on the chain length.

The investigators visualize the chain termination as a multistage process which includes the reversible desorption of the active end of the molecule (or its segments) from the catalyst surface, diffusion of the whole molecule into the bulk, and the monomolecular chain termination taking place in the bulk. The diffusion stage is characterized by a diffusion constant, depending on the chain length. The value of the constant of the termination reaction proper is not affected by the chain length.

The effective termination constant is determined from the formula:

$$k_t = B/[n^{1/2}(n^{1/2} + A)]$$

where A and B are constants, and n is the degree of polymerization.

When the MWD was calculated, it was assumed that polymerization is of steady-state nature, that the activity of catalyst surface does not change with time, and that monomer concentration on the catalyst surface and k_p are constant with time.

Experimental findings were verified using various initiating systems, including Phillips catalytic systems[123–124]. The theoretically calculated integral MWD function is in agreement with experimental data. The model, however, overestimates to a considerable extent the content of both the low- and high-molecular-weight fractions.

In[122] two hypotheses explaining the formation of polyethylene with a wide MWD on Ziegler-Natta catalysts are compared. One hypothesis is the dependence of k_t on chain length, the other, on the heterogeneity of active centers.

Theoretical analysis of polymerization in the presence of a chain-transfer agent has shown that, according to the first hypothesis, $\overline{M}_w/\overline{M}_n$ must decrease sharply and asymptotically to approach 2 with increasing concentration of the chain transfer agent. In the second case, the $\overline{M}_w/\overline{M}_n$ value must remain unchanged.

A comparison of calculated results with the experimental data for ethylene polymerization on $[TiCl_4/(isobutyl)_3Al]$ in the presence of different amounts of hydrogen confirmed the correctness of the first hypothesis.

In the Soviet study[125], a kinetic model was developed for ethylene polymerization in a medium of ethylene chloride on a homogeneous $(C_2H_5)_2TiCl/Al(C_2H_5)Cl_2/Al(C_2H_5)_2Cl$ catalytic system. It was shown that the Ti compound also takes part in the reaction of chain propagation limitation. A distinguishing feature of this catalytic system is the fact that bimolecular chain termination does not play a significant role.

The theoretical scheme took into account the reaction of chain propagation limitation by transfer to the Al-organic compound, monomer, and hydrogen, as well as the chain termination following the monomolecular mechanism (spontaneous termination or the "immurement" of active center).

Calculated and compared with experimental data were the curves in coordinates: polyethylene yield – time, $\overline{P}_w$ – ethylene concentration, $\overline{P}_w$ – Al-organic substance concentration, $\overline{P}_w$ and $\overline{P}_w/\overline{P}_n$ – time. The values of kinetic constants for the individual stages of the process have been specified.

In a series of Soviet studies[126–128] a mathematical simulation was performed to examine the process of stereospecific isoprene polymerization on $TiCl_4/(isobutyl)_3Al$ catalytic system in isopentane solution.

The proposed model takes into account the two stages involved in the initiation of active centers: first, the coordination of isoprene molecules on the catalyst surface, and second, the insertion of coordinated molecules, with the insertion stage being the limiting one (zero order with respect to the monomer). The reactions of chain propagation and transfer also proceed in two stages; the chain propagation limitation reactions include chain transfer to $(isobutyl)_2AlCl$ and hydrogen and chain annihilation due to spontaneous monomolecular termination and termination because of impurities.

From kinetic data the values of elementary constants are estimated at $-20\,°C$. The values are not cited.

The first Soviet investigation on the modeling of MWD in the ionic polymerization of butadiene and isoprene in solution on butyllithium catalyst was published in 1958[127]. In this study one can already find all the elements in the scheme of utilizing MWD to specify the polymerization mechanism.

Theoretical calculations of MWD were performed using direct integration and continuous variables based on the assumption that the catalyst is not associated. Instances where $k_i = k_p$ (no chain limitation) and where there is chain transfer to the monomer are examined.

MWD was experimentally determined by means of rapid sedimentation. For diethyl ether solutions at 30 °C, the relative constants of chain transfer (0.005–0.006 for butadiene and 0.003–0.004 for isoprene) from MWD were estimated.

Later a group of researchers from Leningrad[126] turned to the analysis of stereospecific butadiene polymerization, and in the latest study[129], based on an analysis of their own data and those taken from the literature[130], they made an attempt to create a mathematical model of this process.

Variation of the concentration of active centers (C_C) in Ziegler-Natta catalysts occurred as a result of the fast interaction between the aluminium-organic component (C_A) of the catalyst and its titanium component (C_T). This process is described by the empirical equation

$$C_C = k_1 C_T \left(\frac{C_A}{C_T}\right)^a$$

The addition of monomer molecules to the active center is regarded as a two-stage process (coordination and insertion):

$$R_i + M \underset{k_{-2}}{\overset{k_2}{\rightleftarrows}} R_i M$$

$$R_i M \xrightarrow{k_5} R_{i+1}$$

According to[126], the propagating chains are de- and reactivated in the course of the process:

$$R_i(R_iM) + R_j(R_jM) \underset{k_4}{\overset{k_3}{\rightleftarrows}} P_i(P_iM) + P_j(P_jM)$$

The rate of coordination reaction is observed to be substantially higher than that of insertion and deactivation. Therefore, the following equation holds true throughout the whole process:

$$k_2 C_m \sum_i C_{R_i M} = k_{-2} \sum_i [R_i][M]$$

The concentration of active chains is found from analyzing the total concentration of active centers:

$$\Sigma C_{R_iM} = \frac{k_2 C_m \sum_i C_R}{1 + k_2 C_m}$$

Chain transfer to the aluminium alkyl was also deserved. Using the method of moments the authors obtained an equation for the first three moments of active, temporarily deactivated, and "dead" chains. As a result of a computerized search for the values of constants, based on the model and on the experimental data obtained in a batch reactor (volume = 13 l), some of the values were found to differ considerably from those published in the literature.

The values obtained are not cited in the work.

II. Investigations of Nonisothermal Polymerization Processes

In most of the published studies on the mathematical simulation of polymer synthesis processes, the attention of investigators has been concentrated on calculating the dependencies of molecular weight ($\overline{M}_n$ or $\overline{M}_w$) and MWD on the parameters of the process. Calculations can be performed for a batch or a continuous reactor (of mixing or displacement), or for a cascade of continuous reactors.

In calculations of this kind, two different equations of mixing according to Dankwerts[131] are usually used: one for micromixing and one for complete segregation of flows. The imperfect reactor operation can be taken into account by means of the curve of residence time distribution and various empirical combinations of elementary reactor volumes based on it.

Such an approach is of practical value in analysing the operation of existing industrial reactors. However, when new technological processes are designed, it is necessary to investigate various nonisothermal conditions of synthesis, as well as the effect of the rheology of the system on the peculiarities of its kinetic behavior.

Most of the processes of polymer synthesis are characterized by a high exothermicity combined with a low thermal conductivity coefficient of the monomer-polymer mixture. This relationship between the thermophysical characteristics of the polymerization medium, in combination with the reaction rates observed in actual practice, results in the overheating of the reaction mass and the appearance of a non-uniform distribution of temperature and degree of conversion in space and time.

As a rule, an increase in temperature in the course of polymerization is accompanied, by various kinetic effects. For example, in the radical polymerization of vinyl monomers changes can take place in the concentration of radicals and the time when the gel effect sets in. In addition a process of degradation can be superimposed on the polymerization process. The temperature and conversion non-uniformities occurring in the course of polymerization can change the thermal process itself, converting bulk polymerization into a reaction with propagating front, and vice versa.

Within the framework of this approach, a series of studies conducted by Enikolopian, Davtyan et al.[132–139] dealt with the laws underlying radical polymerization and polycondensation in adiabatic conditions and in conditions favoring a propagating reaction front.

The application of Bodenstein's principle (a quasi-steady state with respect to radicals) to nonisothermal processes is shown not always to be correct.

Numerical calculations using a computer showed the difference between the solutions obtained in steady- and nonsteady-state approximations as depending mainly on the magnitude of $Q/(c\rho)$ (where Q is the reaction thermal effect, c is thermal capacity and ρ is the reaction medium density), activation energy, and pre-exponential factors of initiation (k_i) and termination (k_t) rate constants.

From the equation describing the variations in the concentration of radicals (R·)

$$\frac{d[R\cdot]}{dt} = k_i(t)[I]_t - k_t(t)[R\cdot]^2, \tag{3}$$

and designating the relative deviation of the true $[R\cdot]$ from the quasisteady-state $[\overline{R\cdot}]$ concentration of radicals as $\Delta = [R\cdot]/[\overline{R\cdot}] - 1$, we get:

$$\frac{d\Delta}{dt} = -\Delta(1+x)a(t) + \frac{1+\Delta}{2}\psi(t) \tag{4}$$

where $a(t) = \sqrt{k_i(t)k_t(t)[I]_t}$ $\psi(t) = \frac{1}{RT^2}(E_t - E_i)\frac{dT}{dt} + k_i(t)$

$X = [R\cdot]/[\overline{R\cdot}]$

$[I]_t$ is the current concentration of the initiator.

Analysis of Eqs. (3) and (4) shows that if $a(t) \gg 1$, $\frac{k_t(t)}{a(t)} \ll 1$, then $\Delta \longrightarrow 0$.

In this study the researchers investigate theoretically and experimentally the adiabatic polymerization of styrene, methyl methacrylate, and butyl methacrylate at complex initiation, i.e., when a mixture of initiators with a differing activation energy is used.

A method has been proposed to calculate the MWD in the course of nonisothermal polymerization. The importance of this method lies in reducing the initial kinetic equations to a partial differential equation:

$$\frac{\partial[R(x,t)]}{\partial t} + \frac{\varphi_2(t)}{N} \cdot \frac{\partial[R(x,t)]}{\partial x} = \varphi_1(t)R(x,t) \tag{5}$$

where $x = j/N$, the N number is associated with the chosen scale along the X (or the j) axis, $\varphi_1(t) = k_p(T)[M]$, $\varphi_2(t) = k_p(t)[M] + [k_t(T) + k'_t(T)]\sum_1^\infty[R]$, and $\sum_1^\infty[R]$ is the concentration of macroradicals.

Equation (5) was solved by the method of characteristics.

In[139] theoretical and experimental data are examined for radical polymerization under reaction front propagation conditions, and in[140] the impact of pressure and the gel effect on polymerization kinetics is considered also. In this case, a heat conductivity equation is added to the kinetic equations.

A one-dimensional model, with the constancy of the values of thermophysical constants taken into account, has the form:

$$C\rho \frac{\partial T}{\partial t} = \lambda \frac{\partial^2 T}{\partial x^2} + QW$$

where λ is the heat conductivity coefficient, and W is the reaction rate.

In[139] the researchers studied the nonisothermal kinetics of polyaddition reactions of oxiranes to aromatic amines up to high conversion levels, as well as the kinetic laws governing the curing of epoxide oligomers by diamines under the conditions of a propagating reaction front.

The nonisothermal polymerization process in a tubular reactor at laminar flow is investigated in[141]. Experimental data on the polymerization of styrene under these conditions are presented in[142].

The data on initiated styrene polymerization were used in calculations. The reaction system was regarded as quasi-stationary from a hydrodynamic point of view, meaning that the flow velocity profiles instantaneously adjust themselves to variations of viscosity through temperature and conversion.

Earlier, in[143], this assumption was shown to be correct at the value of Prandtl criterion $Pr = \frac{c\mu(T_0)}{\lambda} \gg 1$ (c is heat capacity; μ is viscosity; λ is the heat conductivity coefficient).

The second assumption was, that flow velocity variation along the radius of the tube is much greater than along its length. Based on these assumptions equations of hydrodynamics can be formulated to describe the radial and the axial velocities of flow.

Heat exchange through the walls was taken into account by Newton's law, and diffusion of substance was disregarded, which is permissible if $Le = \frac{D}{\chi} \ll 1$ (D is the diffusion coefficient, χ the thermal diffusivity).

The mathematical model includes a set of five equations: three partial differential equations (thermal balance, balance with respect to the monomer, and the initiator) and two integral equations determining the magnitude of the radial and axial flow velocities.

According to computerized numerical calculations, polymerization under high-temperature conditions is a non-steady-state process substantially dependent on external heat-removal and heat conduction of the substance.

The effect of conductive heat transfer is described by Frank-Kamentsky's parameter

$$\delta = \frac{r_0^2 EQk_0[M]_0^m[I]_0^n e^{-E/(RT_0)}}{RT_0^2 \lambda}$$

where E and Q are the activation energy and thermal effect of the polymerization reaction, and $k_0[M]_0^m[I]_0^n e^{-E/(RT_0)}$ is the polymerization rate.

At $\delta < 1$ polymerization proceeds under the conditions of self-ignition; i.e., at first a sharp warm-up of the substance is observed in the reactor axial zone. The reaction front is thus formed. Conductive heating of the substance results in the reaction front moving toward the flow of reagents. Stabilization of the reaction front is possible when its position is unchanged.

The profile of the front will become warped due to the distribution of velocities over the radius. Accumulation of high-viscosity reaction products near the walls and the absence of mixing will distort the stabilization of the reaction front. Velocity of the axial flow will increase, and the reaction front will be "expelled" from the reactor.

The quenching and stalling of high-temperature polymerization occurred at all reasonable values of the parameters. As a result, it may be concluded that, in agreement with the data presented in[142], with the laminar flow of reagents in a tubular reactor, it is impossible to achieve polymerization under steady-state high-temperature conditions.

A. Butakov, one of the authors of the study on which the above theoretical calculations are based, published in the same collection of articles a new investigation on the benzoylperoxide initiated bulk polymerization of styrene in laminar flow[144].

Non-uniformity of flow profile in the radial cross-sections of a tubular reactor results either in the "break-down" phenomenon, when at a constant flow rate of reagents the reaction front is expelled from the reactor, or in "choking", when at a constant pressure drop along the reactor length its rupturing is possible.

Butakov found that high-temperature polymerization can proceed under stable conditions if the reactor is constructed in the form of a horizontal elongated spiral. This reactor is characterized by longitudinal temperature profiles with a clearly defined "hot point".

For the first time temperature hysteresis (a multitude of steady-state conditions) has been observed to exist in the homogeneous reaction of a tubular reactor.

Hydrodynamic analysis has shown that the location of the "hot point" on the downward sections of the reactor is associated with the intensive mixing of substances due to the action of oppositely directed inertial and ascential forces.

III. Problems of Designing Industrial Polymerization Reactors

The method of mathematically simulating polymerization process kinetics, MWD, and the structural parameters of products has undoubtedly contributed to our understanding of the nature of processes taking place in industrial equipment. This method makes it possible to solve a number of problems facing technologists and designers. Creating control models suitable for the automation of technological process control is one such goal.

More stringent requirements are imposed on the models in which one can optimize a technological process over a number of parameters.

Proceeding from a kinetic model which includes MWD calculations, to try to substantiate in a strictly logical fashion the "optimum" technological scheme and the "optimum" conditions for a synthesis process is indeed tempting but as yet impossible. In study[76] on modeling the process of thermoinitiated bulk polymerization of styrene, it was convincingly shown how fruitless any attempt is at present to formalize the search for the "optimum" variant.

Before this line of research acquires scientific rigor, much research is needed to correlate the structure and properties of polymers, and to investigate the macrokinetics and hydrodynamics in existing polymerization reactors. At present, however, the empirical choice prevails.

We have already spoken of the shroud of secrecy which obscures the values of effective kinetic constants obtained as a result of modeling concrete technological processes in concrete apparatus. In most cases the reader must be content with the information that the researchers found a satisfactory agreement between their calculations and the results obtained on industrial apparatus.

This situation seems to be caused on the one hand, by the complexity and the diverse nature of the necessary calculations and on the other hand, by the non-fundamental nature of the information obtained.

In a monograph[145], soon to be published by the Khimiya Publishers, an attempt is made to present a critical review of the results achieved in the past 10 years with regard to the design of industrial processes. Although it is still too early to speak of a major breakthrough in the technology of polymer synthesis using mathematical simulation methods, a thorough kinetic analysis makes it possible in some cases to revise the conclusions obtained by technologists.

References

1. Bagdasaryan, H. S.: Teoriya radikal'noy polymerizathii, (Theory of radical polymerizations), Moscow, ANUSSR, 1959
2. Gladishev, G., Gibov, K.: Polymerizathiya priglubokih stepenyah prevraschenya, (Polymerization at high degree of conversion), Alma-ata, Nauka, 1968
3. Bamford, C. H. et al.: The kinetics of vinyl polymerization by radical mechanisms, London, Butterworth, 1958
4. Shaulov, A. Yu. et al.: Europ. Polymer J. *40*, 1077 (1974)
5. Horie, K., Mita, I.: Macromolecules *11*, 1175 (1978)
6. Borgwardt, U., Schnabel, W., Henglein, A.: Makromol. Chem. *127*, 176 (1969)
7. Yasukawa, T., Tacahashi, T., Murakami, K.: J. Chem. Phys. *59*, 3937 (1973)
8. Kozlov, S. V., Ovchinnikov, A. A., Enikolopiyan, N. S.: Vysokomol. Soedin. *A612*, 987 (1970)
9. Szwarc, M.: Carbanions, Living Polymers and Electron-Transfer Processes, New York, John Wiley, 1968
10. Erussalimskii, B. L.: J. polymer. poljarnich monomerov, Leningrad, 1970
11. Berlin, A. A., Korolev, G. V., Kefeli, T. Ya.: Polyesterakrylaty, Moskwa, Chimiya, 1967
12. Garina, E. S. et al.: J. Polym. Sci., Polymer Chem. Ed. *16*, 2199 (1978)
13. Markevich, A. M. et al.: Vysokomol. Soedin. *A17*, 2528 (1975)
14. Kaplan, A. M. et al.: Dokl. Akad. Nauk SSSR *224*, 829 (1975)
15. Olenin, A. V. et al.: Vysokomol. Soedin. *B18*, 219 (1976)
16. Olenin, A. V. et al.: Vysokomol. Soedin. *A20*, 407 (1978)
17. Stoyachenko, I. L. et al.: Vysokomol. Soedin. *A15*, 1899 (1973)
18. Golubev, V. B. et al.: J. Polym. Sci., Polym. Chem. Ed. *2*, 2463 (1973)
19. Kim, L. F. et al.: Internat. Symp. Makromol. Chem., Tashkent, USSR *2*, 52 (1978)

20. Flory, P. J.: J. Am. Chem. Soc. *59*, 241 (1937)
21. Breitenbach, J. W., Springer, A., Abrahamczik, E.: Monatsh. Chem. *41*, 182 (1938)
22. Kamenskaja, S. N., Medvedev, S. S.: Z. Phis. Khim. *14*, 929 (1940)
23. Mayo, F. R.: J. Am. Chem. Soc. *65*, 2324 (1943)
24. Bamford, C. H., White, E. F. T.: Trans. Faraday. Soc. *52*, 715 (1958)
25. Enikolopian, N. S.: J. Polym. Sci. *58*, 1301 (1962)
26. Enikolopian, N. S.: Proc. Nat. Conf. Phys. Coord. Chem. Porphyrine, USSR, Ivanovo, 1979
27. Bamford, C. H., Jenkins, A. D., Johnston, R.: Proc. Roy. Soc. *A-239*, 214 (1957)
28. Dainton, F. S., Collinson, E.: Nature *67*, 853 (1963)
29. Dainton, F. S., Ivin, K. J.: Quart. Rev. *12*, 61 (1958)
30. Berlin, Al. Al., Enikolopian, N. S.: Vysokomol. Soedin, *A-21*, 2671 (1969)
31. Enikolopian, N. S. et al.: J. Polym. Sci. *C16*, 2453 (1967)
32. Kravchik, I. P., Enikolopian, N. S.: Vysokomol. Soedin, *A-19*, 2114 (1967)
33. Berlin, Al. Al. et al.: Dokl. Akad. Nauk, SSSR *184*, 1128 (1969)
34. Berlin, Al. Al. et al.: Vysokomol. Soedin. *A-15*, 554 (1973)
35. Ivanov, V. V., Sabirova, R. D., Enikolopian, N. S.: Dokl. Akad. Nauk, SSSR *183*, 1335 (1968)
36. Berlin, Al. Al. et al.: Vysokomol. Soedin., *A-17*, 643 (1975)
37. Bogdanova, K. A. et al.: Dokl. Akad. Nauk, SSSR 211, 874 (1973)
38. Vorobieva, G. A. et al.: Dokl. Akad. Nauk, SSSR *214*, 273 (1974)
39. Bogdanova, K. A. et al.: Vysokomol. Soedin. *A-17*, 658 (1975)
40. Berlin, Al. Al., Enikolopian, N. S.: Dokl. Akad. Nauk, SSSR *196*, 1111 (1971)
41. Karyuhina, G. A. et al.: ibid. *195*, 1147 (1970)
42. Vorobieva, G. A. et al.: Vysokomol. Soedin. *A-16*, 1493 (1974)
43. Kazanskii, K. S., Entelis, C. G.: in Uspehi Khimii i Phisiki polimerov (Adv. Chem. Phys. Polym.), Moscow Khimiy, 324, 1970
44. Kazanskii, K. S., Solovyanov, A. A., Entelis, S. G.: Europ. Polymer J. *7*, 1421 (1971)
45. Kazanskii, K. S., Solovyanov, A. A., Dubrobsky, S. A.: Makromol. Chem. *179*, 969 (1978)
46. Matyjaszewski, K., Kubisa, P., Penczek, S.: J. Polym. Sci. Polymer Chem. Ed. *12 (6)*, 1333 (1974)
47. Boileau, S., Guerin, P., Sigwalt, P.: Europ. Polym. J. *7*, 1581 (1971)
48. Rasvodovskii, E. F. et al.: J. Macromol. Sci., Chem. *8 (2)*, 241 (1974)
49. Rasvadovskii, E. F. et al.: Dokl. Akad. Nauk, SSSR *198*, 894 (1971)
50. Rasvodovskii, E. F. et al.: Vysokomol. Soedin., *A 15*, 2219 (1973)
51. Rasvodovskii, E. F. et al.: Vysokomol. Soedin. *A-15*, 2233 (1973)
52. Morosovs, I. S. et al.: Dokl. Akad. Nauk, SSSR *209*, 153 (1973)
53. Penchev, P. E., Minin, V. A., Ivanov, V. V.: Polymer *15*, 573 (1974)
54. Enikolopian, N. S. et al.: Vysokomol. Soedin. *A-19*, 1924 (1977)
55. Ivanov, V. V. et al.: ibid. *B-19*, 743 (1972)
56. Morosova, I. S. et al.: Dokl. Akad. Nauk, USSR *215*, 641 (1974)
57. Penchev, P. I. et al.: J. Polym. Sci. Polym. Chem. Ed. *12*, 1881 (1974)
58. Volfson, S. A., Minin, V. A., Shpichinezkaya, L. S.: Dokl. Akad. Nauk, SSSR *234*, 1365 (1977)
59. Berlin, Al. Al., Volfson, S. A., Oleinik, E. F.: Vysokomol. Soedin. *12*, 443 (1970)
60. Shpichinezkaya, L. S. et al.: Plastmassy *11*, 19 (1969)
61. Enikolopian, N. S., Volfson, S. M.: Khimiya i Tehnologiya Polyformaldegida (Chemistry and Technology of Polyformaldehyde), Moscow, Khimiya, 1968
62. USSR Author's certificate 208938; Bulleten isobreteny (USSR Patent) Bull. izobret. *4* (1965)
63. Berlin, Al. Al., Enikolopian, N. S.: Vysokomol. Soedin., *A-10*, 1475 (1968)
64. Berlin, Al. Al. et al.: ibid *A-10*, 1496 (1968)
65. Minin, V. A. et al.: ibid. *A-14*, 3 (1972)
66. Ivanova, L. L. et al.: ibid., *A-17*, 1229 (1975)
67. Minin, V. A.: Dissertation for Candidate of Sciences, Inst. Chem. Phys. USSR, Acad. Sci. Moscow, 1978
68. Jaaks, V., Franzman, G.: Makromol. Chem. *143*, 283 (1971)

69. Jaaks, V., Mathes, N.: ibid. *136*, 295 (1970)
70. Jamasita, J., Ito, K., Nakahira, F.: ibid. *127*, 292 (1969)
71. Enikolopian, N. S., Markevich, M. A., Kochetov, E. V.: Problemy kinetiki elementarnyh khimicheskih reactiy (Problems of kinetics of elementary chemical reactions) Moscow, Nauka, 1973
72. Ranagaes, F. et al.: Dokl. Akad. Nauk, SSSR *200*, 634 (1972)
73. Ranagaes, F. et al.: ibid, *202*, 642 (1972)
74. Chappear, D. C., Simou, R. H.: Adv. Chem. Ser., *91*, 1 (1969)
75. Shinnar, R., Katz, S.: ibid. *109*, 56 (1972)
76. Fan, L. T., Shastry, J. S.: Macromol. Rev. (USA) *7*, 155 (1973)
77. Katz, S., Saidel, G. M.: Am. Inst. Chem. Eng. J. *13*, 319 (1967)
78. Suzuki, T., Forestry, J.: Sci. Japan *48*, 436 (1966)
79. Tsuchida, E., Mimashi, S.: J. Polymer Sci. *A 3*, 1401 (1965)
80. Reinhardt, M., Thiele, R.: Plast. Kautschuk *19*, 9 (1972)
81. Duerksen, J. H., Hamielec, A. E.: J. Polym. Sci., Part *C25*, 155 (1968)
82. Hui, A. W. T., Hamielec, A. E.: ibid., *25*, 167 (1968)
83. Hui, A. W. T., Hamielec, A. E.: Ind. Eng. Chem. Proc., Des. Devl. *8*, 105 (1960)
84. Pzyor, W. A., Coco, J. H.: Macromolecules *3*, 500 (1970)
85. Hui, A. W. T., Hamielec, A. E.: J. Appl. Polymer Sci. *16*, 749 (1972)
86. Volfson, S. A. et al.: Dokl. Akad. Nauk, SSSR *226*, 1355 (1976)
87. Volfson, S. A. et al.: Plastmassy *1*, 9 (1977)
88. Husain, A. S., Hamielec, A. E.: J. Appl. Polym. Sci. *22*, 1207 (1978)
89. Kuchanov, S. I., Bort, D. N.: Vysokomol. Soedin., *A-15*, 2393 (1973)
90. Abdel-Alim, A. H., Hamielec, A. E.: J. Appl. Polym. Sci. *16*, 783 (1972)
91. Kafarov, V. V., Dorohov, I. N., Bulle, H.: Dokl. Akad. Nauk, SSSR *230*, 1402 (1976)
92. Marinin, V. G. et al.: Proc. USSR Conf. simulation of polymerization reactors, Vladimir, p. 46, 1979
93. Balke, S. T., Hamielec, A. E.: J. Appl. Polym. Sci. *17*, 905 (1973)
94. Cardenas, J. N., O'Driscoll, K. F.: J. Polym. Sci., Polym. Chem. Ed. *14*, 66, 8839 (1976)
95. Cardenas, J. N., O'Driscoll, K.F.: ibid. *15*, 1883 (1977)
96. Cardenas, J. N., O'Driscoll, K. F.: ibid. *15*, 2097 (1977)
97. Bamford, C. H., Tompa, H.: Trans. Faraday Soc. *50*, 1097 (1954)
98. Saidel, G. M., Katz, S.: J. Polymer. Sci., *A-2,6*, 1149 (1968)
99. Zeman, R., Amundson, N. R.: Chem. Eng. Sci. *20*, 637 (1965)
100. Saito, O., Nagasubramanian, K., Graisley, W. W.: J. Polymer Sci., *A-2, 7*, 1937 (1969)
101. Malyshev, V. A. et al.: Proc. V. USSR Conf. simulation of chem. processes and reactors (Khimreactor-5), Ufa, *3*, 34 (1974)
102. Thies, J., Schoenemann, K.: Adv. Chem. Ser. *109*, 86 (1972)
103. Hans, T.: Chimia, *28*, 377 (1974)
104. Gutin, B. L. et al.: Polymerizatzionnie prossessy. Apparaturnoe oformlenie i mamematicheskoe modelirovanie (Polymerizational Processes), Leningrad, ONPO "Plastpolymer", p. 54 (1976)
105. Stein, D. J.: Makromol. Chem. *76*, 157 (1964)
106. Stein, D. J.: ibid. *76*, 170 (1964)
107. Graessley, W. W., Mittelhauser, H. M.: J. Polym. Sci. *A-2, 5*, 431 (1967)
108. Graessley, W. H., Hartung, B. D., Uy, W. C.: J. Polym. Sci. *A-2, 7*, 1919 (1969)
109. Chatterjee, A., Kabra, K., Graessley, W. W.: J. Appl. Polym. Sci. *21*, 1751 (1977)
110. Gutin, B. L., Beliaev, B. M.: see Ref. 92, p. 32
111. Copolymerization (ed. G. Ham), Interscience, New York – London – Sydney, 1965
112. Manaresh, P., Passalaaua, V., Pilati, F.: Polymer *16*, 520 (1975)
113. Volfson, S. A., Oshmyan, V. G.: Dokl. Akad. Nauk SSSR *245*, 1424 (1979)
114. Sakurada, J., Higashimura, T., Okamura, S.: J. Polym. Sci. *33*, 496 (1958)
115. Taganov, N. G., Korovina, G. V., Entelis, S. G.: Vysokomol. Soedin. *B-17*, 1, 57 (1975)
116. Taganov, N. G., Korovina, G. V., Entelis, S. G.: ibid. *A-20*, 6, 1393 (1978)
117. Taganov, N. G.: Gel-pronikauschaia hromatografia (gel-chromatography), Inst. Chem. Phys., Chernogolivka, p. 106 (1974)

118. Hitrova, R. A., Proskurina, N. P., Kirchevskaia I. U.: see ref. 92, p. 83
119. Chien, J. G. W.: J. Am. Chem. Soc. *81*, 86 (1959)
120. Chien, J. G. W.: J. Polym. Sci. *A 1*, 1839 (1963)
121. Gordon, M., Roe, R. J.: Polym. *2*, 41 (1961)
122. Roe, R. J.: ibid. *2*, 60 (1961)
123. Wesslau, H.: Makromol. Chem. *20*, 111 (1956)
124. Tung, L. H.: J. Polym. Sci. *20*, 495 (1956)
125. Agasaryan, A. V., Belov, G. P., Davtyan, S. P.: see ref. 92, p. 108
126. Bresler, L. S. et al.: Vysokomol. Soedin. *11-A*, 1165 (1968)
127. Bresler, S. E. et al.: Zh. Tehnicheskoi Physiki (J. Techn. Physics) *28*, 114 (1958)
128. Zak, A. B. et al.: see ref. 92, p. 96
129. Lavrov, V. A. et al.: see ref. 92, p. 88
130. Harwart, M. et al.: Plast.-Kautschuk *22*, 233 (1975)
131. Danckwerts, P. V.: Chem. Eng. Sci. *8*, 93 (1956)
132. Tonoian, A. et al.: Vysokomol. Soedin. *B 16*, 799 (1974)
133. Tonoian, A. et al.: Vysokomol. Soedin. *A 15*, 1847 (1973)
134. Gukasova, E. et al.: Dokl. Akad. Nauk SSSR *231*, 1392 (1976)
135. Tonoian, A. et al.: Thesis Nat. Conf. Proc. Apparat. Production Polym., Moscow, p. 1392, 1977
136. Aleksanyan, G. et al.: Vysokomol. Soedin., *A 17*, 913 (1975)
137. Makarova, S. et al.: ibid. *B-19*, 726 (1977)
138. Davtian, S. et al.: Dokl. Akad. Nauk SSSR *232*, 379 (1977)
139. Arutunian, N. et al.: ibid. Dokl. Akad. Nauk SSSR *214*, 832 (1974)
140. Arutunian, H. et al.: Vysokomol. Soedin. *A 17*, 2115 (1974)
141. Zhurkov, P. V., Bostandian, S. A., Boiazchenko, V. I.: see ref. 92, p. 56
142. Butakov, A. A., Zanin, A. M.: Phys. gorenia i vzriva, *5*, 70 (1978)
143. Bostandianian, S. A., Merzhanov, A. G., Pruchkina, N. M.: jurnal prikl. mechan. i tehnich. fisiki *5*, 30 (1968)
144. See ref. 92, p. 146
145. Volfson, S. A., Enikolopian, N. S.: Rascheti effectivnih polymerizatzionnih processov (Calculation of effective polymerizational processes), Moscow, Khimia, 1980

Received December 5, 1979
W. Kern (editor)

Synthesis, Characterization and Morphology of Poly(butadiene-g-Styrene)

Joseph P. Kennedy and J. M. Delvaux*

Institute of Polymer Science
The University of Akron USA – Akron, OH 44325

This paper concerns the synthesis of poly(butadiene-g-styrene) by cationic grafting of polystyrene cations onto polybutadiene using alkyl halide/alkylaluminum initiating systems and an investigation into the effect of reaction conditions on grafting and mechanism studies with model reactions. Graft characterization by selective oxidative backbone destruction followed by surviving polystyrene branch analysis and morphology studies are also described.

Table of Contents

I. Synopsis 143

II. Introduction 143

III. Experimental 145
    1. Materials and General Conditions 145
    2. Preparation of 1-Chloro-1-(4-methylphenyl)ethane . . 145
    3. Model Reactions 145
    4. Characterization of Products Obtained in Reactions . 145
    5. Grafting Experiments 146
    6. Characterization of Poly(budadiene-g-styrene) . . . 146
    7. Branch Characterization 146
    8. Techniques Used in Morphology Studies 146

IV. Results and Discussion 147
    A. Model Studies 147
        1. Modeling Interactions Between t-$Bu^{\oplus}$ and 1,2- and 1,4-Units in Polybutadiene 148
        2. Modeling the Interaction Between the Growing Polystyrene Cation and 1,2- and 1,4-Units in Polybutadiene 150
        3. Conclusions from Model Experiments 152

---

* Present address: Union Carbide Co., Bound Brook, N. J.

B. Effect of Reaction Conditions on Grafting 153
   1. Effect of Polybutadiene Concentration on Styrene Polymerization Rate 153
   2. Effect of Styrene Conversion on Grafting Efficiency 154
   3. Effect of Temperature on Grafting Efficiency 155
   4. Influence of Medium Polarity on Grafting Efficiency 155
   5. Effect of Polybutadiene Microstructure on Grafting Efficiency . . 155
   6. Effect of Styrene Conversion on Graft Homogeneity and Molecular Weight Distribution 156
C. Characterization of Polystyrene Branches in Poly(butadiene-g-Styrene) 157
   1. Oxidative Degradation of Polybutadiene Backbone 158
   2. Comparison of the Molecular Weights of the Surviving Polystyrene and Homopolystyrene 158
D. Morphology of Poly(butadiene-g-Styrene) 159
E. Conclusions 161

References 162

I. Synopsis

Poly(butadiene-g-styrene) was synthesized by grafting growing polystyrene cations onto polybutadiene. Grafting polystyrene from slightly chlorinated polybutadiene invariably yielded crosslinked products. The mechanism of grafting-onto was investigated by model experiments using 3-hexene and 3-ethyl-1-pentene as models for the 1,2- and 1,4-unsaturations in polybutadiene, and 1-chloro-1-(4-methylphenyl)-ethane as a model for the growing polystyrene cation. The reactivity of the initiating systems, t-butyl chloride/Et_2AlCl and 1-chloro-1-(4-methylphenyl)ethane/Et_2AlCl towards 1,2- and 1,4-unsaturations in polybutadiene were investigated. Tert-butylation of unsaturations in polybutadiene is followed by rapid hydration, however, some proton elimination may also occur. Similarly, the grafting of growing polystyrene cations onto polybutadiene was mimicked by the addition of the 4-methylphenylethane cation to these models. The effect of reaction conditions on graft copolymer synthesis was studied. Temperature, medium polarity, polybutadiene concentration and microstructure affect grafting efficiency. The molecular weight distribution of the graft copolymer broadens with increasing styrene conversion which was attributed to the formation of branchy polystyrene branches. Grafted polystyrene branches were isolated by selective oxidative degradation of the polybutadiene backbone. The molecular weights of grafted polystyrene and homopolystyrene obtained at the same polystyrene conversion level were found to be different. The morphology of poly(butadiene-g-styrene)'s having different overall compositions has been studied by transmission electron microscopy. According to electron micrographs polybutadiene and polystyrene domains are incompatible and form randomly arranged glassy and rubbery domains. Polystyrene domain arrangements are independent of composition in the range studies and the shape of the domains are irregular irrespective of graft composition.

II. Introduction

Graft copolymers can be readily synthesized by cationic "grafting from" or "grafting onto" techniques[1]: Grafting from occurs if an active site generated along a polymer backbone starts to propagate monomer in the system and thus produces branches; grafting onto occurs if a growing polymer chain attacks another polymer and thereby a branch is attached to a preformed backbone[1]. Kennedy et al.[2-9] have studied in great detail the mechanism of initiation and termination of the grafting-from reaction. In particular, the elucidation of the mechanism of initiation of cationic olefin polymerization by alkyl halide/alkylaluminum systems set the basis for highly efficient cationic grafting and a thorough analysis of available cationic grafting data led to the conclusion that grafting from must yield higher grafting efficiencies, G. E., than grafting onto[1]. Several authors[10-15] reported the synthesis of graft copolymers by cationic grafting onto. In particular, Sigwalt et al.[15] described the grafting of polystyrene and polyindene cations onto unsaturated rubbers, i.e., cis-1,4-polybutadiene and EPDM using alkyl halide/Et_2AlCl or $TiCl_4$ initiator systems. According to these authors, grafting may occur by addition of the growing chains to the unsaturated sites of the backbone (grafting onto), or by initiation from the backbone after alkylation of the backbone (grafting from). Initiation by abstraction of an allylic hydrogen by the initiator/coinitiator system as envisioned by Kennedy[16] has also been proposed. Few details were given with regard to the synthesis of poly-(butadiene-g-styrene) mainly because gelation prevented characterization.

This investigation started as a continuation of research into aspects of grafting from[1]. Our original intention was to prepare thermoplastic elastomers by grafting polystyrene branches from lightly chlorinated polybutadiene backbones in conjunction with alkylaluminum coinitiators:

Polybutadiene ⌒⌒⌒ –CHCl– + Et$_2$AlCl ⟶

[Polybutadiene ⌒⌒⌒ –CH$^\oplus$– Et$_2$AlCl$_2^\ominus$] $\xrightarrow{\text{St}}$

Polybutadiene ⌒⌒⌒ –CH–
                     |
                     CH$_2$–CH$^\oplus$ Et$_2$AlCl$_2^\ominus$
                             |
                             C$_6$H$_5$

↓

Poly(butadiene–g–styrene)

Chemistry similar to this has often been used for the preparation of numerous grafts[1]. Surprisingly, we soon found that this usually successful grafting from route invariably yielded gel at about 10% styrene conversion. Inspite of numerous attempts (changing concentrations, solvents composition, temperature etc.) soluble grafts could not be obtained[17]. It is proposed that crosslinking is unavoidable in this system even in the presence of quite large amounts of styrene because the growing polystyrene branches attack the unsaturated polybutadiene backbone leading to gel[17].

These disappointing preliminary studies, however, pointed the way toward a more promising alternate synthesis route. If growing polystyrene cations readily attack polybutadiene, we reasoned, poly(butadiene-g-styrene) may be prepared by initiating styrene polymerization with a small carbocation e.g., t-Bu$^\oplus$, in the presence of polybutadiene, in other words, by grafting polystyrene *onto* polybutadiene. Since the double bond in styrene is more reactive than unsaturations in polybutadiene, chances for propagation of styrene polymerization prior to attack on the potential backbone appeared reasonably high. With this objective in mind, experiments have been carried out for the preparation of poly(butadiene-g-styrene) by the grafting onto technique.

This paper concerns the synthesis of poly(butadiene-g-styrene) by cationic grafting of polystyrene cations onto polybutadiene using alkyl halide/alkylaluminum initiating systems and an investigation into the effect of reaction conditions on grafting and mechanism studies with model reactions. Graft characterization by selective oxidative backbone destruction followed by surviving polystyrene branch analysis and morphology studies are also described.

III. Experimental

1. Materials and General Conditions

All solvents and monomers were purified and stored under nitrogen atmosphere. The t-butyl chloride (t-BuCl, Eastman) was distilled from calcium hydride under nitrogen atmosphere prior to use. Diethylaluminum chloride (Et_2AlCl, Texas Alkyls Co.) was purified by vacuum distillation (bp °C/mm Hg: 110/25). Ethyl chloride (EtCl, Linde Division, Union Carbide Co.) was purified by passing through a column packed with Molecular Sieves and barium oxide. The cis-3-hexene (Chemical Samples Company) was distilled under nitrogen prior to use. The polybutadiene, Diene 35, from the Firestone Tire and Rubber Company and high 1,2-polybutadiene were purified by reprecipitations from benzene into excess acetone.

Reactions were carried out under a dry nitrogen atmosphere in a stainless steel dry box. The moisture level in the dry box, determined by a MEECO Instrument Company Electrolytic Water Analyzer, was kept below 20 ppm. Glassware was dried at 150 °C overnight prior to use and transferred immediately to the dry box.

2. Preparation of 1-Chloro-1-(4-methylphenyl)ethane

1-Chloro-1-(4-methylphenyl)ethane was prepared by reacting 1-hydroxy-1-(4-methylphenyl)ethane (Chemical Samples Co.) with thionyl chloride (Fisher Reagent). The filtrate was distilled under reduced pressure to yield the pure compound.

3. Model Reactions

Reactions were conducted in test tubes fitted with rubber septums caps. The solvent mixture consisted of n-pentane and ethyl chloride (60/40, v/v). Solutions containing the model compound were thermo-equlibrated at 48 °C. The relative concentration of model compounds is as follows: cis-3-hexene 0.3 M, 3-ethyl-1-pentene: 0.3 M, 1-chloro-1-(4-methylphenyl)ethane: 0.02 M. Initiation was effected by adding a suitable volume of Et_2AlCl followed by t-BuCl or 1-chloro-1-(4-methylphenyl)ethane. The concentration of t-BuCl was 0.02 M and that of Et_2AlCl, 0.1 M. After 15 min, reactions were terminated by addition of 1.5 ml methanol. A known amount of t-butylbenzene, 10^{-4} M, used as an internal standard for GC analysis was added. The product was concentrated by evaporation of the lower boiling components.

4. Characterization of Products Obtained in Reactions

Products were characterized using a Varian 2700 Gas Chromatograph coupled with a DuPont 21-490 B Mass Spectrometer. The data were obtained through a Hewlett-Packard 2100 A computer. Molecular weights of eluted fractions were determined using a MC Chromalytics Mass Chromatograph. Details have been described[17].

5. Grafting Experiments

Specific volumes of stock solutions of polybutadiene (5% w/v) in n-heptane were introduced in test tubes or 500 ml three neck flasks charged with monomer and solvent (n-heptane and ethyl chloride) and thermo-equilibrated. Aliquots of Et_2AlCl followed by t-BuCl were added. Reactions were terminated by addition of methanol. Polymers were recovered by precipitation in methanol, filtered and dried. Grafts and homopolymers were separated by selective solvent extraction using acetone and n-pentane. Pure grafts containing 15–60 wt% polystyrene were obtained.

6. Characterization of Poly(butadiene-g-styrene)

Products were characterized by a Varian T-60 H^1 NMR spectrometer, a Waters Associates Ana-Prep Gel Permeation Chromatograph, and a Mecrolab 503 High Speed Membrane Osmometer.

7. Branch Characterization

Degradation of polybutadiene backbones was carried out by epoxidizing the graft copolymer in benzene with m-chlorobenzoic acid followed by hydrolyzing the resulting epoxides and cleaving the glycols with periodic acid[18]. m-Chloroperbenzoic acid (5–10% stoichiometric excess of backbone unsaturation) was added to pure graft dissolved in benzene and the solution was kept for 24 h at room temperature. The oxodized polymer was precipitated with methanol and neutralized with 5% aqueous sodium bicarbonate. The residue was redissolved in tetrahydrofuran, THF, and a saturated aqueous solution of periodic acid, HIO_4, 1.5 times the quantity theoretically required to produce cleavage of the epoxides was added. The mixture was stored for 24 h at room temperature and then, warmed for 1 h. After cooling, methanol was added to precipitate the remaining polystyrene and its molecular weight was determined by GPC.

8. Techniques Used in Morphology Studies

The morphology of poly(butadiene-g-styrene) was studied using a JEM 120 U-Japan Electron Optics Laboratory Co. Ltd. transmission electron microscope operating at 120 KV. Ultra-thin microtomed sections of solvent cast films were prepared at −110 to −150 °C to minimize disruption of sample morphology during sectioning. Contrast between the two phases was achieved by selective staining with osmium tetroxide vapor of polybutadiene domains using ultra-thin sections mounted on a 400 mesh grid[19]. Specimens with an average of 500 Å thickness were found satisfactory for transmission microscopy[20–23]. Samples were annealed at 120 °C for 7 days under vacuum before staining to permit observation of near equilibrium morphology. Samples were microtomed in directions both perpendicular and parallel to the plane of solvent cast films[24, 25].

IV. Results and Discussion

As intimated in the Introduction, our results fall into four areas: 1) model studies whose aim was to provide guidance for subsequent grafting experiments, 2) grafting studies, in particular the elucidation of the effect of reaction variables on grafting details, 3) graft characterization by selective oxidative degradation of the polybutadiene trunk followed by polystyrene branch fragment analysis and 4) morphology studies by transmission electron microscopy.

A. Model Studies

Principle of Model Experiments

The model compounds selected to study the mechanism of grafting styrene onto polybutadiene are shown in Fig. 1. The 1,4- and 1,2-units in polybutadiene were simulated by 3-hexene and 3-ethyl-1-pentene, respectively. The reactivity difference between the 1,4- and 1,2-units in polybutadiene was studied.

The reactivity of the t-BuCl/Et$_2$AlCl initiating system toward 1,2- and 1,4- unsaturations was investigated. Another set of experiments was designed to simulate the reactivity of growing polystyrene cations toward 1,2- and 1,4-units in the backbone. The 1-chloro-1-(4-methylphenyl)ethane/Et$_2$AlCl system was selected to simulate the reactive end group of the growing polystyrene cation.

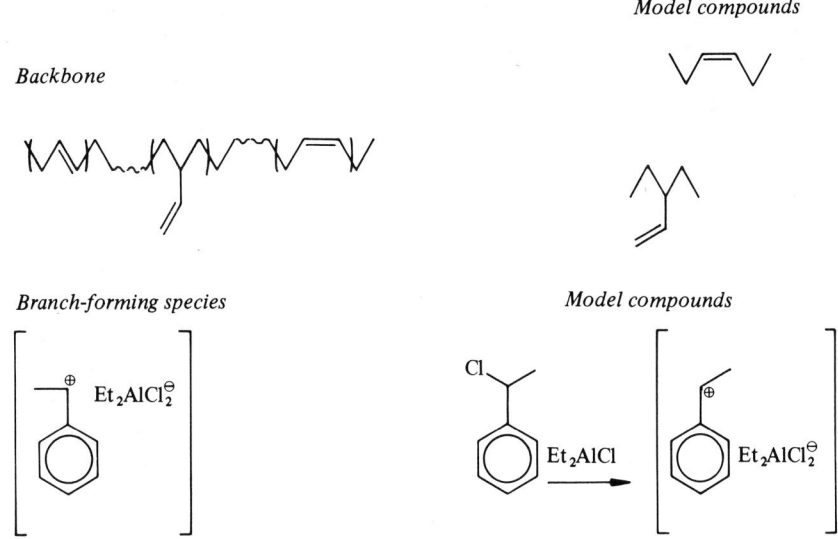

Fig. 1. Models used to investigate the grafting of PSt$^\oplus$ onto PBd

1. Modeling Interactions Between t-Bu$^\oplus$ and 1,2- and 1,4-Units in Polybutadiene

1.1. The t-BuCl/Et$_2$AlCl Plus 3-Ethyl-1-pentene Reaction

The aim of this study was to elucidate the reaction between the t-BuCl/Et$_2$AlCl initiating system and pendent vinyl groups (1,2-units) in polybutadiene. The t-BuCl was added to charges of 3-ethyl-1-pentene and Et$_2$AlCl in n-pentane/ethyl chloride solution at -48 °C. The products were analyzed by gas chromatography and the molecular weights of the products (characterized by GC peaks) were determined by mass spectroscopy and mass chromatography.

Besides traces of residual ethyl chloride and n-pentane, the presence of three compounds was detected: C_7H_{14}, identified to be the 3-ethyl-1-pentene starting material, $C_{11}H_{24}$, identified to be the hydridated add-on product of t-Bu$^\oplus$ to 3-ethyl-1-pentene (and isomers), and $C_{18}H_{38}$, identified as the product obtained by addition of $C_{11}H_{23}^\oplus$ to 3-ethyl-1-pentene followed by hydridation (and isomers). Analysis of the products formed in the reaction of t-BuCl/Et$_2$AlCl with 3-ethyl-1-pentene indicates that termination occurs exclusively by hydridation and that ethylation is

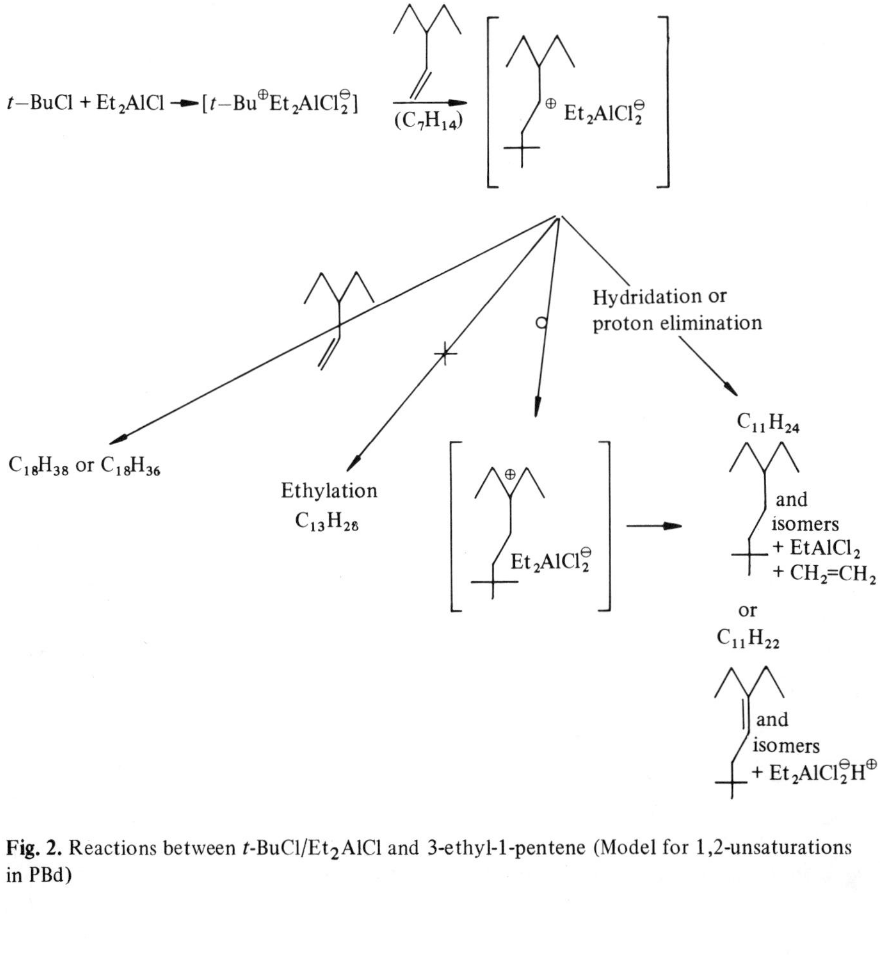

Fig. 2. Reactions between t-BuCl/Et$_2$AlCl and 3-ethyl-1-pentene (Model for 1,2-unsaturations in PBd)

absent. Proton elimination cannot be ruled out, even though products corresponding to this event could not be detected due to a lack of sufficiently accurate analytical method[17]. Surprisingly there was no evidence for termination by ethylation, i.e., t-butylation of 3-ethyl-1-pentene followed by ethylation leading to $C_{13}H_{28}$. The overall scheme of the reactions occurring between the t-BuCl/Et$_2$AlCl initiating system and 3-ethyl-1-pentene, simulating 1,2-units in polybutadiene, is shown in Fig. 2.

1.2. The t-BuCl/Et$_2$AlCl Plus Cis-3-hexene Reaction

The reactivity of t-Bu$^\oplus$ toward 1,4-units in polybutadiene was studied using the t-BuCl/Et$_2$AlCl initiating system and cis-3-hexene as model for 1,4-enchainment in polybutadiene. The analytical methods were the same as those described in 1.1[17].

Besides traces of residual solvent, the presence of three types of compounds was detected. In addition to unreacted cis-3-hexene (C_6H_{12}), the most important product was $C_{10}H_{22}$ or $C_{10}H_{20}$ which most likely was formed by t-butylation of cis-3-hexene followed by rapid hydridation. Addition of t-Bu$^\oplus$ to cis-3-hexene produces a secondary carbenium ion which may rearrange to form a tertiary carbenium ion prior to hydridation. Significantly, the absence of $C_{12}H_{26}$ indicates the absence of ethylation and suggest addition of t-Bu$^\oplus$ to 1,4-unsaturations followed by rapid termination by hydridation. Proton elimination might have also occurred but could not be definitively established.

The overall scheme of the reactions occurring between the t-BuCl/Et$_2$AlCl initiating system and cis-3-hexene, simulating 1,4-units in polybutadiene, is presented in Fig. 3.

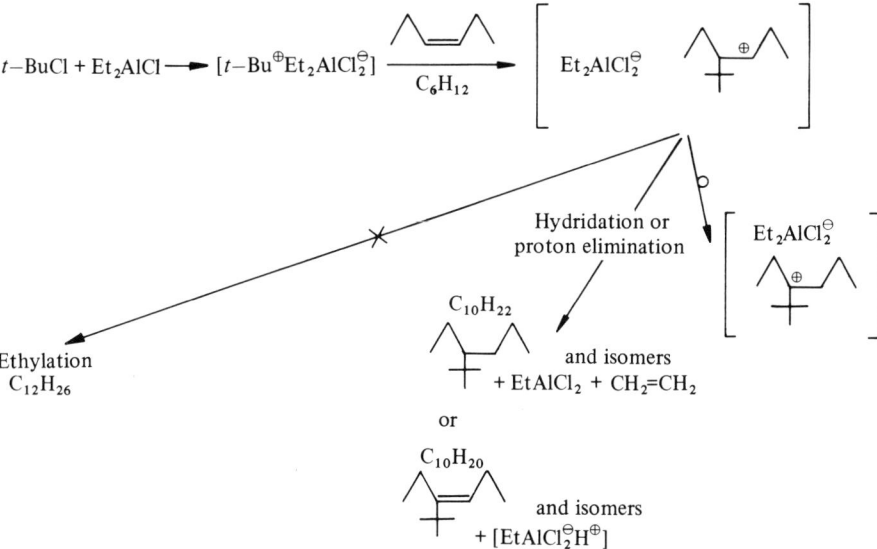

Fig. 3. Reactions between t-BuCl/Et$_2$AlCl and cis-3-hexene (Model for 1,4-unsaturation in PBd)

2. Modeling the Interaction Between the Growing Polystyrene Cation and 1,2- and 1,4-Units in Polybutadiene

2.1. The 1-Chloro-1-(4-methylphenyl)ethane/Et$_2$AlCl Plus 3-Ethyl-1-pentene Reaction

The reactivity of growing polystyrene cations toward 1,2-units in polybutadiene has been studied by treating 3-ethyl-1-pentene with the 1-chloro-1-(4-methylphenyl)-ethane/Et$_2$AlCl system. Addition of Et$_2$AlCl to 1-chloro-1-(4-methylphenyl)ethane gives rise to an ion pair which simulates the growing polystyrene carbocation.

Numerous products were identified by GC coupled with mass chromatography and mass spectroscopy. In addition to *n*-pentane, C_5H_{12}, and residual 3-ethyl-1-pentene, C_7H_{14}, the presence of C_9H_{12} or C_9H_{10} and $C_{11}H_{16}$, i.e., 1-ethyl-4-methyl-benzene, 1-vinyl-4-methylbenzene and 1-methyl-1-(4-methylphenyl)propane have been identified. These compounds most likely arise by hydridation and ethylation, respectively, of the carbenium ion formed in the reaction of Et$_2$AlCl and 1-chloro-1-(4-methylphenyl)ethane. Proton elimination may also occur, however, the analytical information was insufficient for a firm conclusion in this regard. The major product (plus minor isomers) is $C_{16}H_{26}$ or $C_{16}H_{24}$, formed by addition of the 4-ethylphenylethane cation to 3-ethyl-1-pentene followed by hydridation or proton elimination. Significantly, products corresponding to the addition of the 4-methylphenylethane cation to 3-ethyl-1-pentene followed by ethylation, $C_{18}H_{30}$, were not detect-

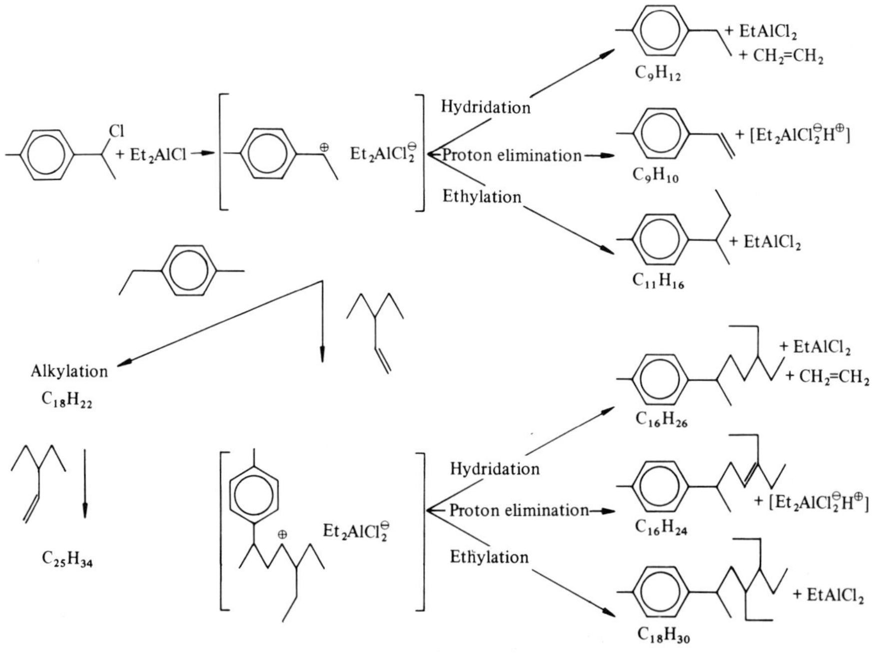

Fig. 4. Reactions between 1-chloro-1-(4-methylphenyl)ethane/Et$_2$AlCl and 3-ethyl-1-pentene (Model for 1,2-unsaturations in PBd)

ed. Small quantities of other products corresponding to alkylation or dimerization have also been detected and isolated. The formers probably arise via ring alkylation by 4-methylphenylethane cations of aromatic compounds. The presence of $C_{14}H_{30}$ was associated with the dimer of 3-ethyl-1-pentene,

On the basis of this information, the overall scheme of the reaction of 1-chloro-1-(4-methylphenyl)ethane/Et_2AlCl system with 3-ethyl-1-pentene is shown in Fig. 4.

2.2. The 1-Chloro-1-(4-methylphenyl)ethane/Et_2AlCl Plus *Cis*-3-hexene Reaction

The reactivity of the growing polystyrene cation toward 1,4-units in polybutadiene was explored with model experiments using 1-chloro-1-(4-methylphenyl)ethane/-Et_2AlCl plus *cis*-3-hexene. 1-Chloro-1-(4-methylphenyl)ethane in conjunction with Et_2AlCl produces a carbocation which may be regarded as a model for the growing polystyrene cation.

Species C_9H_{12} and $C_{11}H_{16}$ were assigned to the products of the reaction occurring between 1-chloro-1-(4-methylphenyl) ethane and Et_2AlCl followed by hydridation and ethylation, respectively. The main product was $C_{15}H_{22}$ which suggests the addition of the 4-methylphenylethane cation to *cis*-3-hexene followed by hydridation or proton elimination. Significantly, the absence of a $C_{17}H_{28}$ species indicates the absence of termination by ethylation.

The overall scheme constructed from this information is shown in Fig. 5.

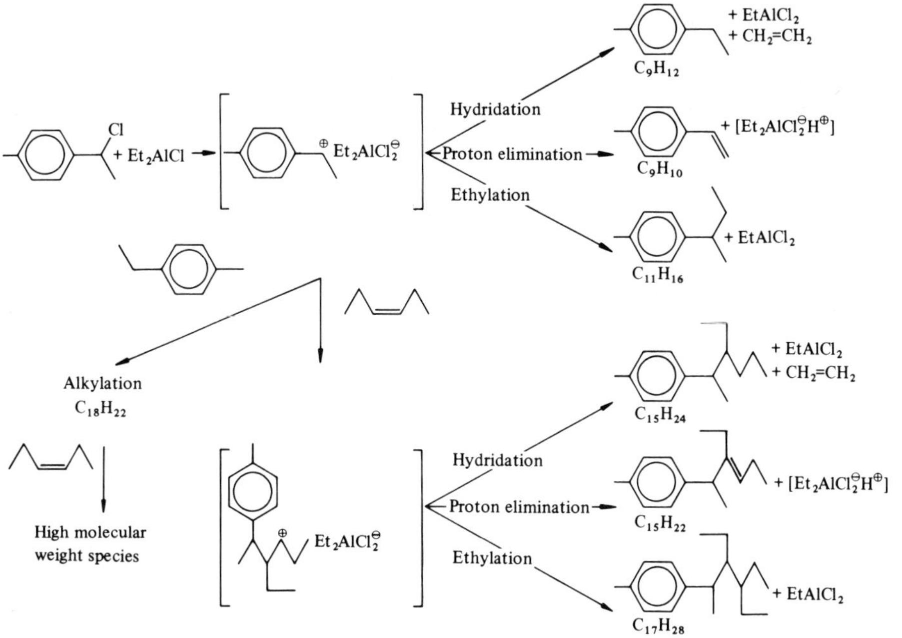

Fig. 5. Reactions between 1-chloro-1-(4-methylphenyl)ethane/Et_2AlCl with *cis*-3-hexene (Model for 1,4-unsaturations in PBd)

3. Conclusions from Model Experiments

According to the results of model experiments, the t-BuCl/Et$_2$AlCl and CH$_3$C$_6$H$_4$CH(CH$_3$)Cl/Et$_2$AlCl systems generate the t-Bu$^\oplus$ Et$_2$AlCl$_2^\ominus$ and CH$_3$C$_6$H$_4^\oplus$CHCH$_3$Et$_2$AlCl$_2^\ominus$ ion pairs, respectively, and the addition of these cations to 1,2- or 1,4-unsaturations in polybutadiene is rapidly followed by hydridation or proton elimination, whereas ethylation by Et$_2$AlCl$_2^\ominus$ is much less likely to occur. Evidently, steric compression around the macrocation formed by the addition of t-Bu$^\oplus$ or CH$_3$C$_6$H$_4^\oplus$CHCH$_3$ to polybutadiene prevents the approach of the Et$_2$AlCl$_2^\ominus$ counter anion and thus prevents ethylation. The above cations mimick, respectively, the growing polyisobutylene and polystyrene carbocations. Thus it is conceivable that soluble poly(butadiene-g-styrene) can be obtained by grafting onto (see preliminary experiments mentioned in the Introduction) because the attack of growing polystyrene cation on polybutadiene is rapidly followed by hydridation-termination, i.e., the macrocation formed upon polystyrene cation addition to polybutadiene is flanked by three polymer sequences so that it is sterically too compressed to sustain propagation or other reactions leading to crosslinked product. In contrast, the macrocation formed by grafting from is flanked by only two polymer sequences, can add monomer, and thus ultimately will result in crosslinked product.

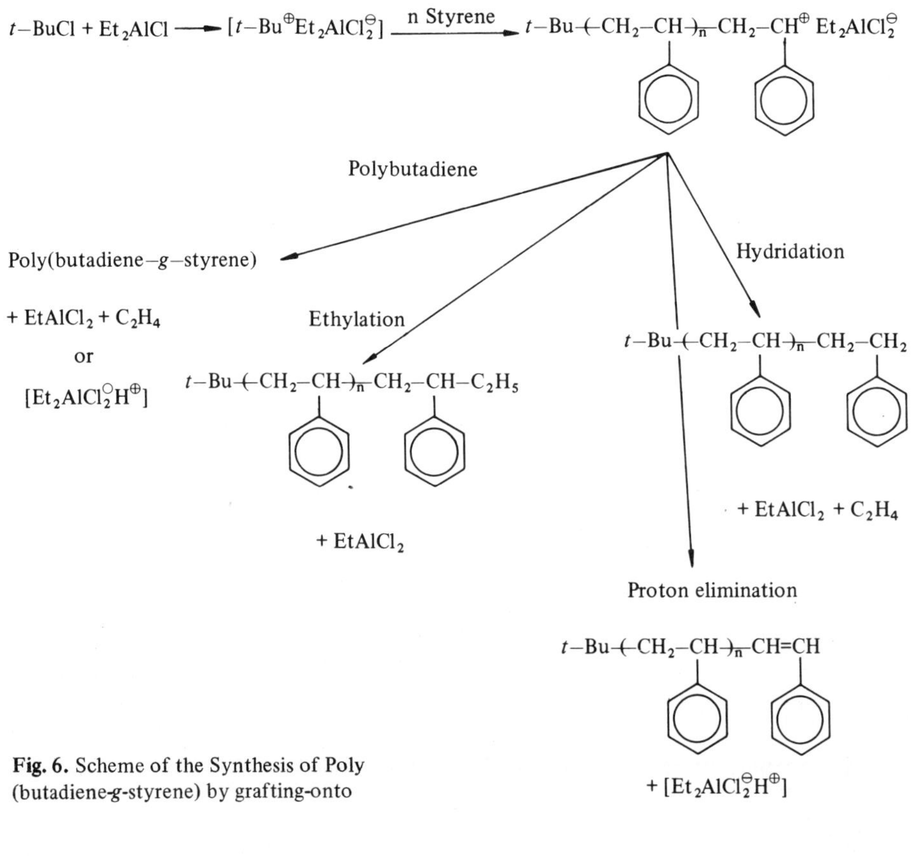

Fig. 6. Scheme of the Synthesis of Poly(butadiene-g-styrene) by grafting-onto

Figure 6 summarizes reactions anticipated to occur during the grafting of growing polystyrene cations onto polybutadiene: The growing polystyrene cation chain may terminate by hydridation, proton elimination, ethylation and alkylation of polybutadiene; on the basis of model experiments the latter event is expected to be followed by rapid hydridation to yield poly(butadiene-g-styrene).

Rapid hydridation also follows t-butylation of polybutadiene, i.e., a reaction which mimicks the attack of growing polyisobutylene cations on polybutadiene. Since our aim was the investigation of reactions yielding thermoplastic elastomers, i.e., combination of rubbery and glossy sequences, experimentation toward poly-(butadiene-g-isobutylene) syntheses has not been carried out during these investigations.

B. Effect of Reaction Conditions on Grafting

Results of scouting grafting onto experiments indicated that soluble poly(butadiene-g-styrene)'s can readily and quite efficiently be obtained by effecting styrene polymerization with the t-BuCl/Et$_2$AlCl initiating system in the presence of various polybutadienes. Thus it appeared of interest to follow up this lead and to investigate the mechanism of this technique, in particular to elucidate the effect of experimental conditions on grafting details.

1. Effect of Polybutadiene Concentration on Styrene Polymerization Rate

Figure 7 shows the results of experiments designed to determine the effect of polybutadiene concentration on the rate of grafting. As the concentration of polybuta-

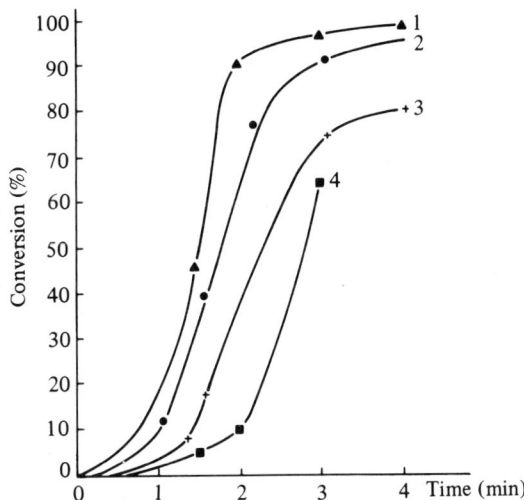

Fig. 7. Effect of polybutadiene concentration on styrene polymerization rate ([Et$_2$AlCl] = $1.5 \cdot 10^{-2}$ M, [t-BuCl] = $0.3 \cdot 10^{-2}$ M, Al/Cl = 5, [styrene] = 0.46 M, -30 °C, n-heptane/EtCl = 60/40). 1: [polybutadiene] = 0 g/l, 2: [polybutadiene] = 6.95 g/l, 3: [polybutadiene] = 13.92 g/l, 4: [polybutadiene] = 18.30 g/l

diene increases, the rate of styrene conversion decreases. Evidently the backbone acts as a retarder which is expected to occur in the presence of a grafting onto mechanism. Increasing polybutadiene concentration also lengthens the induction period. In contrast, in grafting from experiments (not shown), conversion increased with increasing backbone concentration since in this case, the backbone functions as initiator. According to model experiments t-Bu$^\oplus$ adds to backbone unsaturations which reduces the concentration of initiating carbocations. This effect becomes increasingly important with polybutadiene concentration.

2. Effect of Styrene Conversion on Grafting Efficiency

According to the findings shown in Fig. 8, G. E. increases with styrene conversion. These results are in agreement with the proposition that polystyrene is grafted onto polybutadiene. The low G. E. obtained at low styrene conversions indicates that only a few growing polystyrene chains attack polybutadiene. With increasing styrene conversion the probability for a growing polystyrene cation to encounter polybutadiene also increases and thus G. E. increases. Possibly, grafted polystyrene branches prevent growing polystyrene cations from approaching the backbone. Similar observations have been made by Nakamura et al.[26] who speculated that steric hindrance created by branches was responsible for a decrease in G. E. at high conversion during the grafting of polystyrene onto poly(vinyl-p-nitrobenzoate) by free radical technique.

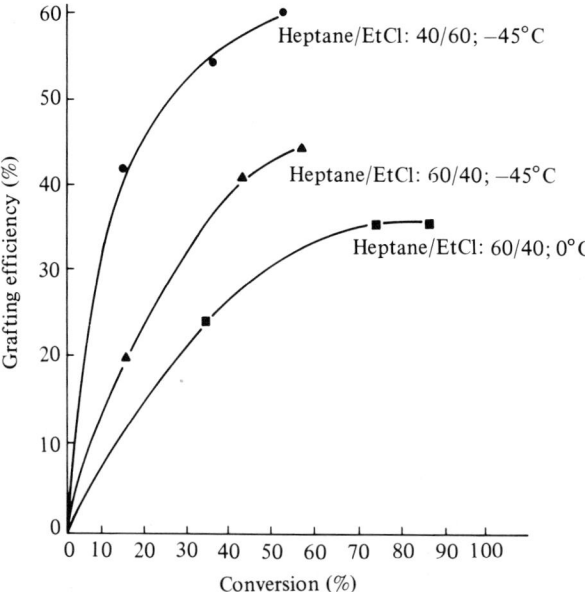

Fig. 8. Effect of temperature and medium polarity on grafting efficiency. ● [Et$_2$AlCl] = 5×10^{-3} M, [t-BuCl] = 10^{-3} M, [polybutadiene ∆] = 15 g/l; -45 °C; ▲ [Et$_2$AlCl] = $20 \cdot 10^{-2}$ M, [t-BuCl] = $40 \cdot 10^{-3}$ M, [polybutadiene] = 11.6 g/l; -45 °C; ■ [Et$_2$AlCl] = 10^{-2} M, [t-BuCl] = $4 \cdot 10^{-3}$ M, [polybutadiene] = 11.5 g/l; 0 °C, [styrene] = 0.46 M

3. Effect of Temperature on Grafting Efficiency

The effect of temperature on G. E. has been studied. Decreasing temperature from 0 °C to −45 °C results in increasing G. E. as shown in Fig. 8. Assuming that initiator/coinitiator concentration has little effect on G. E., these results could be explained by assuming reduction of chain transfer to monomer with decreasing temperature.

4. Influence of Medium Polarity on Grafting Efficiency

Data on the effect of n-heptane/EtCl concentration on G. E. are also shown in Fig. 8. The initiator/coinitiator concentration in the higher polarity experiment (n-heptane/EtCl, 40/60) was reduced to obtain a slower controlled reaction. Evidently, grafting efficiency increases with increasing medium polarity. This effect could be explained by assuming enhanced separation of carbenium ion/counter ion pairs, i.e., increased solvation. Grafting occurs when a growing polystyrene chain adds to polybutadiene followed by rapid termination. Increased G. E. indicates that this addition, like propagation, is strongly affected by medium polarity.

5. Effect of Polybutadiene Microstructure on Grafting Efficiency

The effect of polybutadiene microstructure on G. E. was studied using a high 1,2-polybutadiene synthesized by anionic polymerization and a commercial product, Diene 35, which contains mainly 1,4-unsaturations. According to the data shown in Fig. 9, higher G. E.'s have been obtained in the presence of the high 1,4-polybutadiene than with high 1,2-polybutadiene. These results are in agreement with those obtained in model experiments. The reaction of t-BuCl/Et_2AlCl and cis-3-hexene, model for 1,4-unsaturation, yields predominantly the addition product of

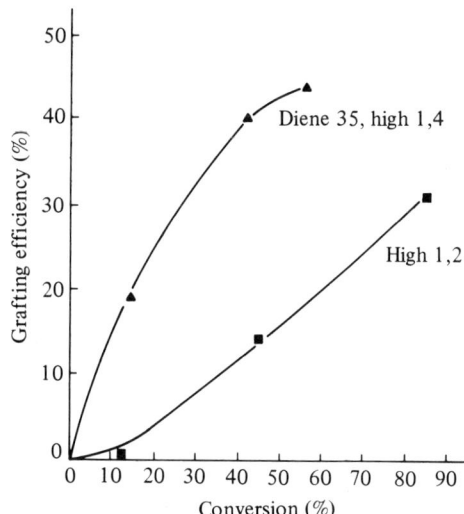

Fig. 9. Effect of polybutadiene microstructure on grafting efficiency ([Et_2AlCl] = $2 \cdot 10^{-2}$ M, t-BuCl = $4 \cdot 10^{-3}$ M, [styrene] = 0.46 M, [polybutadiene] = 11.6 g/l, n-heptane/EtCl = 60/40, −45 °C

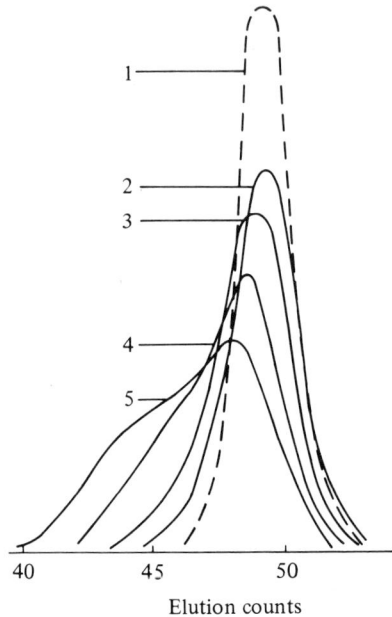

Fig. 10. Normalized gel permeation chromatograms of poly(butadiene-*g*-styrenes) of various compositions. Polystyrene wt. % = 1 : 15.8, 2 : 27, 3 : 41.5, 4 : 53.3, 5 : 55.3

t-Bu$^{\oplus}$ plus *cis*-3-hexene. With 3-ethyl-1-pentene, model for 1,2-unsaturation, the same reaction leads to various addition products. The reaction of 1-chloro-1-(4-methylphenyl)ethane/Et$_2$AlCl, model for growing polystyrene cation with *cis*-3-hexene, yields mostly the addition product.

6. Effect of Styrene Conversion on Graft Homogeneity and Molecular Weight Distribution

The effect of styrene conversion on poly(butadiene-*g*-styrene) homogeneity and on homopolystyrene molecular weight distribution during grafting has been studied by GPC. Traces of pure poly(butadiene-*g*-styrene)'s of different compositions are shown in Fig. 10.

The molecular weight distribution of the grafts broadens with increasing polystyrene content in the graft. The appearance of a shoulder at high molecular weights indicates a change in the structure of the grafts with increasing styrene conversions. According to model experiments it is likely that growing polystyrene cations alkylate grafted polystyrene chains, particularly at increasingly high styrene conversions. Overberger et al.[27] showed that styrene polymerization leads to branching at high conversion and attributed this to ring alkylation. Alkylation of polystyrene branches by growing polystyrene cation may occur at about 50% polystyrene in the graft.

The homopolystyrene formed during grafting has been recovered by selective solvent (acetone) extraction and characterized by GPC. According to the data shown in Table 1, the average molecular weight of homopolystyrene decreases with increasing polystyrene content in the graft of styrene conversion. As the medium becomes poorer in styrene, it is expected that the polystyrene chains become shorter. Longer

Table 1. Molecular weights of graft copolymers and homopolystyrene formed during graft synthesis

Styrene conversion[a] %	Polystyrene[a] Wt.%	Graft[b] $\bar{M}_n \, 10^{-4}$	Homopolystyrene[c] $\bar{M}_w \, 10^{-3}$	$\bar{M}_n \, 10^{-3}$
9.2	15.8	10.1	32.8	19.0
28	27	13.8	27.8	15.9
47.2	41.5	16.8	28.2	14.5
69.1	53.3	21.6	25.8	13.4
79.3	55.5	37.9	23.3	12.0

[a] Gravimetry; [b] Osmometry; [c] GPC

chains terminate by either alkylating the polybutadiene backbone or the pendant phenyl groups of the grafted polystyrene chains.

In the presence of appreciable amounts of polystyrene branches, the polybutadiene backbone may be shielded by the branches. Under these conditions preferential ring alkylation, i.e., branchy branch formation may occur. Further data to corroborate this assumption are presented below.

C. Characterization of Polystyrene Branches in Poly(butadiene-g-Styrene)

An important objective of this research was to determine directly the molecular weights and the molecular weight distribution of polystyrene branches in poly(butadiene-g-styrene).

The choice of a polybutadiene backbone allows selective oxidative destruction of the unsaturated trunk, thus branch characterization independent of and undisturbed by the backbone.

Independent branch analysis of graft copolymers obtained by cationic grafting from has been performed[18]. Methods available for the determination of olefinic unsaturations have been reviewed by Muller[28]. Barnard[29] modified Boer and Kooyman's method[30] and degraded copolymers of natural rubber and methyl methacrylate or styrene, however, under these conditions, partial degradation of the saturated branches was observed. OsO$_4$ was used by Kolthoff et al.[31] to determine the polystyrene content in styrene-butadiene copolymers. Permanganate oxidation of double bonds gives 1,2-diols and the addition of periodate converts the diols to carbonyl compounds. Blanchette and Nielsen[32] used this method for the degradation of GR-S rubbers crosslinked with polystyrene. Recently Minoura et al.[33] have applied this method to isolate selectively polystyrene branches of a natural rubber-polystyrene graft copolymer. The surviving polystyrene branches were characterized by intrinsic viscosity measurements. The similarity between the molecular weights of surviving polystyrene and of homopolystyrene led these authors to conclude that the rubber had been totally degraded. The number and length of the grafted polystyrene branches were also determined.

1. Oxidative Degradation of Polybutadiene Backbone

The *m*-chloro perbenzoic/periodic system was selected to degrade the polybutadiene backbone. Addition of *m*-chloro perbenzoic acid to polybutadiene followed by hydrolysis results in the formation of a glycol which is further converted to cyclic ester by HIO_4 addition[34-36]. Progress of degradation was followed by infrared spectroscopy. Representative infrared spectra are shown in Fig. 11. The surviving polystyrene exhibits the same infrared characteristics as a homopolystyrene. The presence of an absorption band at 5.85 microns associated with the carbonyl vibration and the absence of an absorption at 10.25 microns associated with =C–H out of plane deformation vibration indicate complete backbone degradation and the presence of only surviving polystyrene.

2. Comparison of the Molecular Weights of the Surviving Polystyrene and Homopolystyrene

The molecular weight and molecular weight distribution, MWD, of surviving polystyrene fragments have been determined by GPC and compared with those obtained for homopolystyrene recovered by acetone extraction in the same grafting experiment. According to the results shown in Table 2, molecular weights of grafted polystyrene are consistently higher than those of homopolystyrene; however, their molecular weight distributions are similar. The difference in molecular weights may be explained by assuming alkylation of polystyrene chains during the grafting of polystyrene onto polybutadiene. Evidently the growing polystyryl cations not only attack backbone unsaturations, but may also attack pendent phenyl groups of poly-

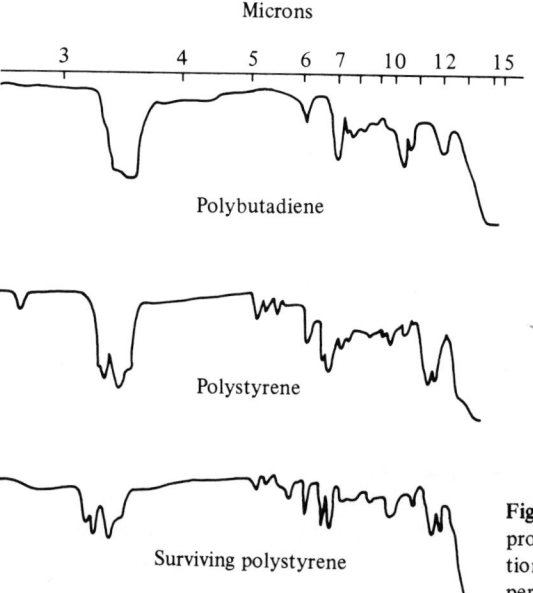

Fig. 11. Infrared spectra of reaction products obtained by oxidative degradation with *m*-chloroperbenzoic acid and periodic acid

Table 2. Polystyrene molecular weights

Styrene (wt.%)	Extracted polystyrene			Grafted polystyrene		
	$\bar{M}_n$ 10^{-3}	$\bar{M}_w$ 10^{-3}	$\bar{M}_w/\bar{M}_n$	$\bar{M}_n$ 10^{-3}	$\bar{M}_w$ 10^{-3}	$\bar{M}_w/\bar{M}_n$
17	16	32	2	24	46.5	1.9
27	17	29	1.7	24	39	1.6
36	13.5	24	1.8	24	43.5	1.8
41.5	14.5	28	1.9	22	42	1.9
47	16	31	1.9	24.5	46.5	1.9
53.3	15	26	1.8	22	39	1.8
54	10.5	20	2	20	38.5	1.9

styrene chains which will subsequently become part of the graft. Results of model experiments according to which the $CH_3-\overset{\oplus}{C}H-C_6H_4-CH_3$ cation attacks not only 1,2- and 1,4-unsaturations but also aromatic rings corroborate this proposition. These results differ from those obtained by grafting styrene from polychloroprene by Kennedy and Metzler[18] who reported identical molecular weights of extracted and grafted polystyrene. The latter observation supports the assumption[37-39] that molecular weights and MWD of grafted branches in grafting from can be estimated from the extracted homopolymer.

The difference in grafted polystyrene and homopolystyrene molecular weights substantiates the proposition of branchy-branch formation. Due to this circumstance it is not possible to determine branch frequencies in the graft copolymer.

D. Morphology of Poly(butadiene-g-Styrene)

While a very large body of morphological information is available on sequential copolymers and particularly on thermoplastic elastomers of the triblock type very little is known of the morphology of graft copolymers synthesized by cationic polymerization techniques. Oziomek and Kennedy[40] have studied the morphology of a graft comprising a commercially available styrene-butadiene-copolymer SBR backbone and polyisobutylene branches containing 65% SBR and 35% PIB. The electron micrograph of this polymer showed two distinct polymer phases, PIB domains dispersed in a continous SBR matrix.

An objective in synthesizing poly(butadiene-g-styrene) by cationic grafting onto was to study the morphology of such grafts by transmission electron microscopy and the effect of composition on graft morphology. The unsaturated nature of the backbone permitted such a study.

Figures 12 through 15 show electron micrographs of samples containing 15.8, 27 and 43 wt.% polystyrene. The micrographs show random arrangement of polystyrene domains. The domains are irregular, irrespective of graft composition. Table 3 lists the estimated size of polystyrene domains as a function of the molecular weight of polystyrene branches.

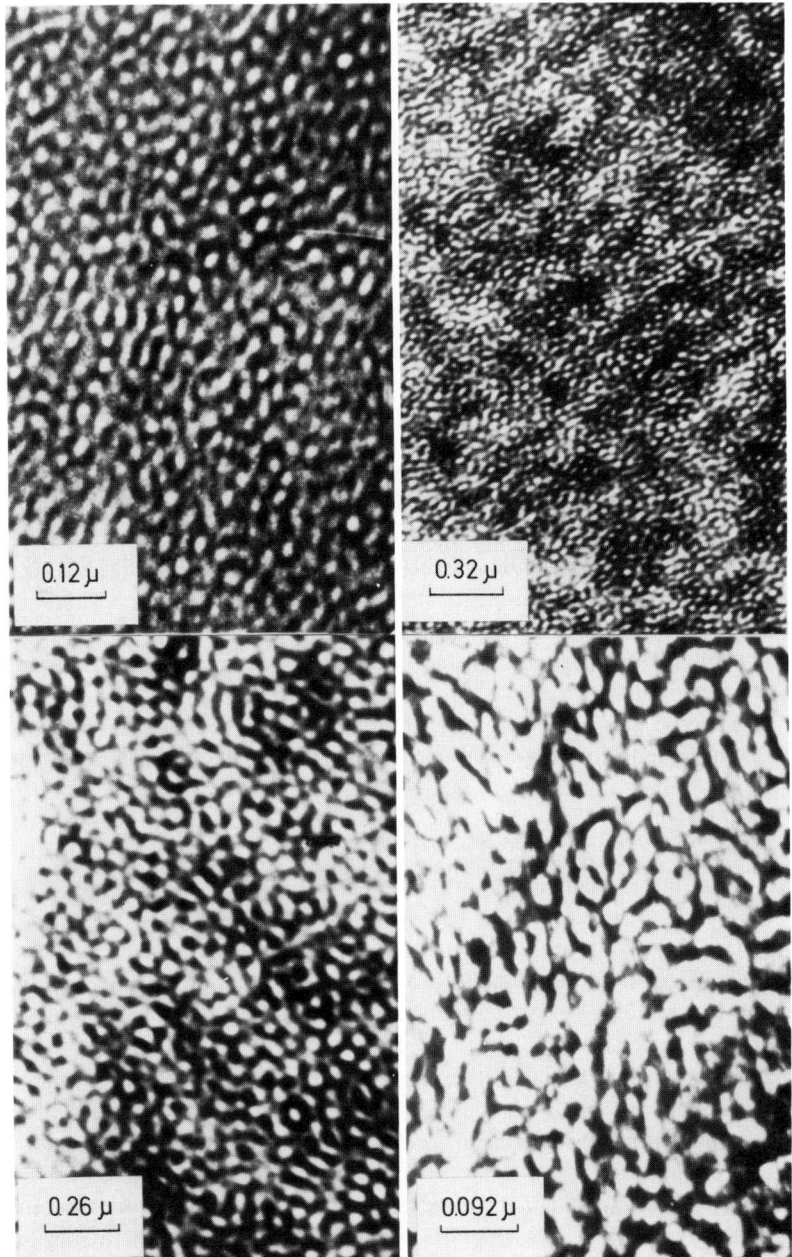

Fig. 12–15. Electron micrographs of poly(butadiene-g-styrene) s. (*12*) PSt = 15.8 wt. %, perpendicular cut (*13*) PSt = 27.0 wt. %, parallel cut (*14*). Same as 2 with perpendicular cut (*15*) PSt = 43.0 wt. %, perpendicular cut

Table 3. Estimated domain size of grafted polystyrene

Polystyrene content wt. % polystyrene[a]	Plane of observation	Estimated average domain diameter[b] Å
15.8	perpendicular	50–180
27	parallel	180–280
27	perpendicular	200–300
43	perpendicular	200–400

[a] By ^{1}H NMR; [b] By electron microscopy

The morphology observed may be due to configurational constraints on the polybutadiene imposed by the grafted polystyrene chains. Domain arrangements appear to be composition independent; i.e., polystyrene seems to form irregular domains over the 15.8 to 43 wt.% polystyrene range. The configurational restraint imposed on the rubbery backbone by the randomly placed relatively broad molecular weight distribution rigid branches may prevent the organization of phases into highly regular domains observed with monodisperse block copolymers.

E. Conclusions

The mechanism of reactions that occur during the polymerization of styrene initiated by t-BuCl/Et$_2$AlCl in the presence of polybutadiene have been studied by model experiments. The t-BuCl/Et$_2$AlCl initiating system most likely gives rise to the t-Bu$^\oplus$/Et$_2$AlCl ion pair that reacts with both 1,2- and 1,4-units in polybutadiene mimicked by 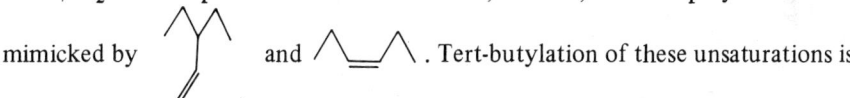 and . Tert-butylation of these unsaturations is followed by rapid hydridation; however, some proton elimination may also occur. The extent of elimination is difficult to assess. The absence of ethylated products corresponding to termination by ethylation suggests that the cation formed by addition of t-Bu$^\oplus$ to either type of unsaturation is sterically hindered. This steric compression is also likely to prevent initiation of styrene polymerization by the polybutadiene cation that is formed by addition of t-Bu$^\oplus$ to backbone unsaturations.

The t-Bu$^\oplus$ formed by reaction of t-BuCl and Et$_2$AlCl also represents a model for growing polyisobutylene cations. According to this study, growing polyisobutylene cations could add to polybutadiene and it is anticipated that soluble poly(butadiene-g-isobutylene) could be synthesized by grafting polyisobutylene onto polybutadiene.

Poly(butadiene-g-styrene) was synthesized by grafting polystyrene onto polybutadiene. Styrene conversions up to 80% corresponding to G. E. of 50% were obtained. The graft copolymers were soluble over the styrene conversion range of 0 to 80%. Temperature, medium polarity, polybutadiene concentration and microstructure affect G. E.

Styrene conversion and polystyrene content in the graft increase with time of synthesis. High G. E.'s were obtained in polar media with high 1,4-polybutadiene. The molecular weight distribution of the graft broadens with increasing styrene conversion. This phenomenon was attributed to formation of branchy-branches in the graft. If termination occurs prior to alkylation of the backbone the resulting homopolystyrene will be of lower average molecular weight that the polystyrene which will be attached to the backbone. Accordingly, the fact that the molecular weight of grafted polystyrene is higher than that of homopolystyrene is in concord with the proposition of a grafting onto process.

Electron micrographs of poly(butadiene-g-styrene) obtained by cationic grafting growing polystyrene cations onto preformed 1,4-polybutadiene exhibits two phases, the arrangement of which is composition independent.

Acknowledgements. Financial support by the National Science Foundation (DMR-77-27618) is gratefully acknowledged.

References

1. *"Cationic Graft Copolymerizations"* J. P. Kennedy (ed.): J. Appl. Polymer Symp. *30* (1977)
2. Reibel, L., Kennedy, J. P., Chung, Y. L.: J. Org. Chem. *42* (1977)
3. Kennedy, J. P.: J. Polym. Sci., A-1, 6, 3139 (1968)
4. Kennedy, J. P.: J. Macromol. Sci. Chem., A 3 (5), 861 (1969)
5. Kennedy, J. P.: J. Org. Chem. *35*, 532 (1970)
6. Kennedy, J. P., Desai, N. V., Sivaram, S.: J. Am. Chem. Soc. *95*, 6386 (1973)
7. Kennedy, J. P., Sivaram, S.: J. Org. Chem. *38*, 2262 (1973)
8. Kennedy, J. P.: J. Macromol. Sci. Chem., A 7 (4), 969 (1973)
9. Kennedy, J. P., Ichikawa, M.: Polym. Eng. Sci. *14*, 322 (1974)
10. Hass, H. C., Kamath, P. M., Schuler, N. W.: J. Polym. Sci. *24*, 85 (1957)
11. Overberger, C. G., Burns, C. M.: J. Polym. Sci. A-1, 7, 333 (1969)
12. Ueno, Y. et al.: J. Polym. Sci., A-1, 5, 339 (1967)
13. Monoura, Y. et al.: J. Polym. Sci., A-1, 4, 1665 (1966)
14. Jaacks, V., Kern, W.: Makromol. Chem. *83*, 71 (1965)
15. Sigwalt, P., Polton, A., Miskovic, M.: J. Polym. Sci., Part C. Polym. Symp. *56*, 13 (1976)
16. Kennedy, J. P.: J. Macromol. Sci., A-6 *2*, 339 (1972)
17. Delvaux, M. J.: Ph. D. thesis, The University of Akron, 1980
18. Kennedy, J. P., Metzler, D. K.: "Cationic Graft Copolymerization," J. P. Kennedy (Ed.) J. Appl. Polymer Symp. *30*, 141 (1977)
19. Kato, K.: J. Polym. Sci., *B-4*, 35 (1966)
20. Matsuo, M., Sagae, S., Asai, H.: Polymer *10*, 79 (1969)
21. Pedemonte, E. et al.: Polymer *14*, 145 (1973)
22. Lewis, P. R., Price, C.: Polymer *13*, 20 (1972)
23. Price, C., Watson, A. G., Chow, M. T.: Polymer *13*, 333 (1972)
24. Lewis, P. R., Price, C.: Nature *223*, 494 (1969)
25. Lewis, P. R., Price, C.: Polymer *12*, 258 (1971)
26. Nakamura, S., Sato, H., Matsuzaki, K.: J. Appl. Polym. Sci., B *11*, 221 (1973); ibid., J. Appl. Polym. Sci. *22*, 2011 (1978)
27. Overberger, C. G., Kamath, P. M., Endres, G. F.: J. Am. Chem. Soc. *84*, 4813 (1962)
28. Muller, K.: "Functional Group Determination of Olefinic and Acetylenic Unsaturation," Academic Press, London, 1974

29. Barnard, D.: J. Polym. Sci. *22*, 213 (1956)
30. Boer, H., Kooyman, E. C.: Anal. Chim. Acta *5*, 550 (1951)
31. Kolthoff, I. M., Lee, T. S., Carr, C. W.: J. Polymer Sci. *1*, 429 (1946)
32. Blanchette, J., Nielsen, L. E.: J. Polym. Sci. *20*, 317 (1956)
33. Mori, V., Minoura, Y., Imoto, M.: Angew. Makromol. Chem. *25*, 1 (1958)
34. Duke, F. R.: J. Am. Chem. Soc. *69*, 3054 (1947)
35. Price, C. C., Knell, M.: J. Am. Chem. Soc. *64*, 552 (1942)
36. Buist, J. D., Bunton, C. A., Miles, J. H.: J. Am. Chem. Soc., 743 (1959)
37. Thame, N. G., Lundberg, R. D., Kennedy, J. P.: J. Polymer Sci., A-1 *10*, 2507 (1972)
38. Kennedy, J. P., Davidson, D. L.: "Cationic Graft Copolymerization," J. P. Kennedy (Ed.) J. Appl. Polymer Symp. *30*, 13; *ibid.*, 51
39. Kennedy, J. P., Vidal, A.: J. Polym. Sci., A-1 *13*, 1765 (1975)
40. Oziomek, J., Kennedy, J. P.: J. Appl. Polymer Symp. *30*, 91 (1977)

Received June 23, 1980
J. P. Kennedy (editor)

Author Index Volumes 1–38

Allegra, G. and *Bassi, I. W.:* Isomorphism in Synthetic Macromolecular Systems. Vol. 6, pp. 549–574.
Andrews, E. H.: Molecular Fracture in Polymers. Vol. 27, pp. 1–66.
Ayrey, G.: The Use of Isotopes in Polymer Analysis. Vol. 6, pp. 128–148.
Baldwin, R. L.: Sedimentation of High Polymers. Vol. 1, pp. 451–511.
Basedow, A M. and *Ebert, K.:* Ultrasonic Degradation of Polymers in Solution. Vol. 22, pp. 83–148.
Batz, H.-G.: Polymeric Drugs. Vol. 23, pp. 25–53.
Bergsma, F. and *Kruissink, Ch. A.:* Ion-Exchange Membranes. Vol. 2, pp. 307–362.
Berlin, Al. Al., Volfson, S. A., and *Enikolopian, N. S.:* Kinetics of Polymerization Processes. Vol. 38, pp. 89–140.
Berry, G. C. and *Fox, T. G.:* The Viscosity of Polymers and Their Concentrated Solutions. Vol. 5, pp. 261–357.
Bevington, J. C.: Isotopic Methods in Polymer Chemistry. Vol. 2, pp. 1–17.
Bird, R. B., Warner, Jr., H. R., and *Evans, D. C.:* Kinetik Theory and Rheology of Dumbbell Suspensions with Brownian Motion. Vol. 8, pp. 1–90.
Biswas, M. and *Maity, C.:* Molecular Sieves as Polymerization Catalysts. Vol. 31, pp. 47–88.
Block, H.: The Nature and Application of Electrical Phenomena in Polymers. Vol. 33, pp. 93–167.
Böhm, L. L., Chmelíř, M., Löhr, G., Schmitt, B. J. und *Schulz, G. V.:* Zustände und Reaktionen des Carbanions bei der anionischen Polymerisation des Styrols. Vol. 9, pp. 1–45.
Bovey, F. A. and *Tiers, G. V. D.:* The High Resolution Nuclear Magnetic Resonance Spectroscopy of Polymers. Vol. 3, pp. 139–195.
Braun, J.-M. and *Guillet, J. E.:* Study of Polymers by Inverse Gas Chromatography. Vol. 21, pp. 107–145.
Breitenbach, J. W., Olaj, O. F. und *Sommer, F.:* Polymerisationsanregung durch Elektrolyse. Vol. 9, pp. 47–227.
Bresler, S. E. and *Kazbekov, E. N.:* Macroradical Reactivity Studied by Electron Spin Resonance. Vol. 3, pp. 688–711.
Bucknall, C. B.: Fracture and Failure of Multiphase Polymers and Polymer Composites. Vol. 27, pp. 121–148.
Bywater, S.: Polymerization Initiated by Lithium and Its Compounds. Vol. 4, pp. 66–110.
Bywater, S.: Preparation and Properties of Star-branched Polymers. Vol. 30, pp. 89–116.
Carrick, W. L.: The Mechanism of Olefin Polymerization by Ziegler-Natta Catalysts. Vol. 12, pp. 65–86.
Casale, A. and *Porter, R. S.:* Mechanical Synthesis of Block and Graft Copolymers. Vol. 17, pp. 1–71.
Cerf, R.: La dynamique des solutions de macromolecules dans un champ de vitesses. Vol. 1, pp. 382–450.
Cesca, S., Priola, A. and *Bruzzone, M.:* Synthesis and Modification of Polymers Containing a System of Conjugated Double Bonds. Vol. 32, pp. 1–67.
Cicchetti, O.: Mechanisms of Oxidative Photodegradation and of UV Stabilization of Polyolefins. Vol. 7, pp. 70–112.
Clark, D. T.: ESCA Applied to Polymers. Vol. 24, pp. 125–188.

Coleman, Jr., L. E. and *Meinhardt, N. A.:* Polymerization Reactions of Vinyl Ketones. Vol. 1, pp. 159–179.
Crescenzi, V.: Some Recent Studies of Polyelectrolyte Solutions. Vol. 5, pp. 358–386.
Davydov, B. E. and *Krentsel, B. A.:* Progress in the Chemistry of Polyconjugated Systems. Vol. 25, pp. 1–46.
Dole, M.: Calorimetric Studies of States and Transitions in Solid High Polymers. Vol. 2, pp. 221–274.
Dreyfuss, P. and *Dreyfuss, M. P.:* Polytetrahydrofuran. Vol. 4, pp. 528–590.
Dušek, K. and *Prins, W.:* Structure and Elasticity of Non-Crystalline Polymer Networks. Vol. 6, pp. 1–102.
Eastham, A. M.: Some Aspects of the Polymerization of Cyclic Ethers. Vol. 2, pp. 18–50.
Ehrlich, P. and *Mortimer, G. A.:* Fundamentals of the Free-Radical Polymerization of Ethylene. Vol. 7, pp. 386–448.
Eisenberg, A.: Ionic Forces in Polymers. Vol. 5, pp. 59–112.
Elias, H.-G., Bareiss, R. und *Watterson, J. G.:* Mittelwerte des Molekulargewichts und anderer Eigenschaften. Vol. 11, pp. 111–204.
Fischer, H.: Freie Radikale während der Polymerisation, nachgewiesen und identifiziert durch Elektronenspinresonanz. Vol. 5, pp. 463–530.
Fujita, H.: Diffusion in Polymer-Diluent Systems. Vol. 3, pp. 1–47.
Funke, W.: Über die Strukturaufklärung vernetzter Makromoleküle, insbesondere vernetzter Polyesterharze, mit chemischen Methoden. Vol. 4, pp. 157–235.
Gal'braikh, L. S. and *Rogovin, Z. A.:* Chemical Transformations of Cellulose. Vol. 14, pp. 87–130.
Gallot, B. R. M.: Preparation and Study of Block Copolymers with Ordered Structures, Vol. 29, pp. 85–156.
Gandini, A.: The Behaviour of Furan Derivatives in Polymerization Reactions. Vol. 25, pp. 47–96.
Gandini, A. and *Cheradame, H.:* Cationic Polymerization. Initiation with Alkenyl Monomers. Vol. 34/35, pp. 1–289.
Gerrens, H.: Kinetik der Emulsionspolymerisation. Vol. 1, pp. 234–328.
Goethals, E. J.: The Formation of Cyclic Oligomers in the Cationic Polymerization of Heterocycles. Vol. 23, pp. 103–130.
Graessley, W. W.: The Etanglement Concept in Polymer Rheology. Vol. 16, pp. 1–179.
Hay, A. S.: Aromatic Polyethers. Vol. 4, pp. 496–527.
Hayakawa, R. and *Wada, Y.:* Piezoelectricity and Related Properties of Polymer Films. Vol. 11, pp. 1–55.
Heitz, W.: Polymeric Reagents. Polymer Design, Scope, and Limitations. Vol. 23, pp. 1–23.
Helfferich, F.: Ionenaustausch. Vol. 1, pp. 329–381.
Hendra, P. J.: Laser-Raman Spectra of Polymers. Vol. 6, pp. 151–169.
Henrici-Olivé, G. und *Olivé, S.:* Kettenübertragung bei der radikalischen Polymerisation. Vol. 2, pp. 496–577.
Henrici-Olivé, G. und *Olivé, S.:* Koordinative Polymerisation an löslichen Übergangsmetall-Katalysatoren. Vol. 6, pp. 421–472.
Henrici-Olivé, G. and *Olivé, S.:* Oligomerization of Ethylene with Soluble Transition-Metal Catalysts. Vol. 15, pp. 1–30.
Henrici-Olivé, G. and *Olivé, S.:* Molecular Interactions and Macroscopic Properties of Polyacrylonitrile and Model Substances. Vol. 32, pp. 123–152.
Hermans, Jr., J., Lohr, D. and *Ferro, D.:* Treatment of the Folding and Unfolding of Protein Molecules in Solution According to a Lattic Model. Vol. 9, pp. 229–283.
Holzmüller, W.: Molecular Mobility, Deformation and Relaxation Processes in Polymers. Vol. 26, pp. 1–62.
Hutchison, J. and *Ledwith, A.:* Photoinitiation of Vinyl Polymerization by Aromatic Carbonyl Compounds. Vol. 14, pp. 49–86.
Iizuka, E.: Properties of Liquid Crystals of Polypeptides: with Stress on the Electromagnetic Orientation. Vol. 20, pp. 79–107.
Ikada, Y.: Characterization of Graft Copolymers. Vol. 29, pp. 47–84.
Imanishi, Y.: Syntheses, Conformation, and Reactions of Cyclic Peptides. Vol. 20, pp. 1–77.

Inagaki, H.: Polymer Separation and Characterization by Thin-Layer Chromatography. Vol. 24, pp. 189–237.
Inoue, S.: Asymmetric Reactions of Synthetic Polypeptides. Vol. 21, pp. 77–106.
Ise, N.: Polymerizations under an Electric Field. Vol. 6, pp. 347–376.
Ise, N.: The Mean Activity Coefficient of Polyelectrolytes in Aqueous Solutions and Its Related Properties. Vol. 7, pp. 536–593.
Isihara, A.: Intramolecular Statistics of a Flexible Chain Molecule. Vol. 7, pp. 449–476.
Isihara, A.: Irreversible Processes in Solutions of Chain Polymers. Vol. 5, pp. 531–567.
Isihara, A. and *Guth, E.:* Theory of Dilute Macromolecular Solutions. Vol. 5, pp. 233–260.
Janeschitz-Kriegl, H.: Flow Birefringence of Elastico-Viscous Polymer Systems. Vol. 6, pp. 170–318.
Jenkins, R. and *Porter, R. S.:* Unpertubed Dimensions of Stereoregular Polymers. Vol. 36, pp. 1–20.
Jenngins, B. R.: Electro-Optic Methods for Characterizing Macromolecules in Dilute Solution. Vol. 22, pp. 61–81.
Kamachi, M.: Influence of Solvent on Free Radical Polymerization of Vinyl Compounds. Vol. 38, pp. 55–87.
Kawabata, S. and *Kawai, H.:* Strain Energy Density Functions of Rubber Vulcanizates from Biaxial Extension. Vol. 24, pp. 89–124.
Kennedy, J. P. and *Chou, T.:* Poly(isobutylene-*co*-β-Pinene): A New Sulfur Vulcanizable, Ozone Resistant Elastomer by Cationic Isomerization Copolymerization. Vol. 21, pp. 1–39.
Kennedy, J. P. and *Delvaux, J. M.:* Synthesis, Characterization and Morphology of Poly(butadiene-g-Styrene). Vol. 38, pp. 141–163.
Kennedy, J. P. and *Gillham, J. K.:* Cationic Polymerization of Olefins with Alkylaluminium Initiators. Vol. 10, pp. 1–33.
Kennedy, J. P. and *Johnston, J. E.:* The Cationic Isomerization Polymerization of 3-Methyl-1-butene and 4-Methyl-1-pentene. Vol. 19, pp. 57–95.
Kennedy, J. P. and *Langer, Jr., A. W.:* Recent Advances in Cationic Polymerization. Vol. 3, pp. 508–580.
Kennedy, J. P. and *Otsu, T.:* Polymerization with Isomerization of Monomer Preceding Propagation. Vol. 7, pp. 369–385.
Kennedy, J. P. and *Rengachary, S.:* Correlation Between Cationic Model and Polymerization Reactions of Olefins. Vol. 14, pp. 1–48.
Kennedy, J. P. and *Trivedi, P. D.:* Cationic Olefin Polymerization Using Alkyl Halide – Alkylaluminum Initiator Systems. I. Reactivity Studies. II. Molecular Weight Studies. Vol. 28, pp. 83–151.
Kissin, Yu. V.: Structures of Copolymers of High Olefins. Vol. 15, pp. 91–155.
Kitagawa, T. and *Miyazawa, T.:* Neutron Scattering and Normal Vibrations of Polymers. Vol. 9, pp. 335–414.
Kitamaru, R. and *Horii, F.:* NMR Approach to the Phase Structure of Linear Polyethylene. Vol. 26., pp. 139–180.
Knappe, W.: Wärmeleitung in Polymeren. Vol. 7, pp. 477–535.
Koningsveld, R.: Preparative and Analytical Aspects of Polymer Fractionation. Vol. 7.
Kovacs, A. J.: Transition vitreuse dans les polymers amorphes. Etude phénoménologique. Vol. 3, pp. 394–507.
Krässig, H. A.: Graft Co-Polymerization of Cellulose and Its Derivatives. Vol. 4, pp. 111–156.
Kraus, G.: Reinforcement of Elastomers by Carbon Black. Vol. 8, pp. 155–237.
Kreutz, W. and *Welte, W.:* A General Theory for the Evaluation of X-Ray Diagrams of Biomembranes and Other Lamellar Systems. Vol. 30, pp. 161–225.
Krimm, S.: Infrared Spectra of High Polymers. Vol. 2, pp. 51–72.
Kuhn, W., Ramel, A., Walters, D. H., Ebner, G. and *Kuhn, H. J.:* The Production of Mechanical Energy from Different Forms of Chemical Energy with Homogeneous and Cross-Striated High Polymer Systems. Vol. 1, pp. 540–592.
Kunitake, T. and *Okahata, Y.:* Catalytic Hydrolysis by Synthetic Polymers. Vol. 20, pp. 159–221.
Kurata, M. and *Stockmayer, W. H.:* Intrinsic Viscosities and Unperturbed Dimensions of Long Chain Molecules. Vol. 3, pp. 196–312.

Ledwith, A. and *Sherrington, D. C.:* Stable Organic Cation Salts: Ion Pair Equilibria and Use in Cationic Polymerization. Vol. 19, pp. 1–56.
Lee, C.-D. S. and *Daly, W. H.:* Mercaptan-Containing Polymers. Vol. 15, pp. 61–90.
Lipatov, Y. S.: Relaxation and Viscoelastic Properties of Heterogeneous Polymeric Compositions. Vol. 22, pp. 1–59.
Lipatov, Y. S.: The Iso-Free-Volume State and Glass Transitions in Amorphous Polymers: New Development of the Theory. Vol. 26, pp. 63–104.
Mano, E. B. and *Coutinho, F. M. B.:* Grafting on Polyamides. Vol. 19, pp. 97–116.
Mengoli, G.: Feasibility of Polymer Film Coating Through Electroinitiated Polymerization in Aqueous Medium. Vol. 33, pp. 1–31.
Meyerhoff, G.: Die viscosimetrische Molekulargewichtsbestimmung von Polymeren. Vol. 3, pp. 59–105.
Millich, F.: Rigid Rods and the Characterization of Polyisocyanides. Vol. 19, pp. 117–141.
Morawetz, H.: Specific Ion Binding by Polyelectrolytes. Vol. 1, pp. 1–34.
Mulvaney, J. E., Oversberger, C. C. and *Schiller, A. M.:* Anionic Polymerization. Vol. 3, pp. 106–138.
Okubo, T. and *Ise, N.:* Synthetic Polyelectrolytes as Models of Nucleic Acids and Esterases. Vol. 25, pp. 135–181.
Osaki, K.: Viscoelastic Properties of Dilute Polymer Solutions. Vol. 12, pp. 1–64.
Oster, G. and *Nishijima, Y.:* Fluorescence Methods in Polymer Science. Vol. 3, pp. 313–331.
Overberger, C. G. and *Moore, J. A.:* Ladder Polymers. Vol. 7, pp. 113–150.
Patat, F., Killmann, E. und *Schiebener, C.:* Die Absorption von Makromolekülen aus Lösung. Vol. 3, pp. 332–393.
Penczek, S., Kubisa, P. and *Matyjaszewski, K.:* Cationic Ring-Opening Polymerization of Heterocyclic Monomers. Vol. 37, pp. 1–149.
Peticolas, W. L.: Inelastic Laser Light Scattering from Biological and Synthetic Polymers. Vol. 9, pp. 285–333.
Pino, P.: Optically Active Addition Polymers. Vol. 4, pp. 393–456.
Plate, N. A. and *Noah, O. V.:* A Theoretical Consideration of the Kinetics and Statistics of Reactions of Functional Groups of Macromolecules. Vol. 31, pp. 133–173.
Plesch, P. H.: The Propagation Rate-Constants in Cationic Polymerisations. Vol. 8, pp. 137–154.
Porod, G.: Anwendung und Ergebnisse der Röntgenkleinwinkelstreuung in festen Hochpolymeren. Vol. 2, pp. 363–400.
Pospíšil, J.: Transformations of Phenolic Antioxidants and the Role of Their Products in the Long-Term Properties of Polyolefins. Vol. 36, pp. 69–133.
Postelnek, W., Coleman, L. E., and *Lovelace, A. M.:* Fluorine-Containing Polymers. I. Fluorinated Vinyl Polymers with Functional Groups, Condensation Polymers, and Styrene Polymers. Vol. 1, pp. 75–113.
Rempp, P., Herz, J., and *Borchard, W.:* Model Networks. Vol. 26, pp. 107–137.
Rigbi, Z.: Reinforcement of Rubber by Carbon Black. Vol. 36, pp. 21–68.
Rogovin, Z. A. and *Gabrielyan, G. A.:* Chemical Modifications of Fibre Forming Polymers and Copolymers of Acrylonitrile. Vol. 25, pp. 97–134.
Roha, M.: Ionic Factors in Steric Control. Vol. 4, pp. 353–392.
Roha, M.: The Chemistry of Coordinate Polymerization of Dienes. Vol. 1, pp. 512–539.
Safford, G. J. and *Naumann, A. W.:* Low Frequency Motions in Polymers as Measured by Neutron Inelastic Scattering. Vol. 5, pp. 1–27.
Schuerch, C.: The Chemical Synthesis and Properties of Polysaccharides of Biomedical Interest. Vol. 10, pp. 173–194.
Schulz, R. C. und *Kaiser, E.:* Synthese und Eigenschaften von optisch aktiven Polymeren. Vol. 4, pp. 236–315.
Seanor, D. A.: Charge Transfer in Polymers. Vol. 4, pp. 317–352.
Seidl, J., Malinský, J., Dušek, K. und *Heitz, W.:* Makroporöse Styrol-Divinylbenzol-Copolymere und ihre Verwendung in der Chromatographie und zur Darstellung von Ionenaustauschern. Vol. 5, pp. 113–213.
Semjonow, V.: Schmelzviskositäten hochpolymerer Stoffe. Vol. 5, pp. 387–450.

Semlyen, J. A.: Ring-Chain Equilibria and the Conformations of Polymer Chains. Vol. 21, pp. 41–75.
Sharkey, W. H.: Polymerizations Through the Carbon-Sulphur Double Bond. Vol. 17, pp. 73–103.
Shimidzu, T.: Cooperative Actions in the Nucleophile-Containing Polymers. Vol. 23, pp. 55–102.
Silvestri, G., Gambino, S., and *Filardo, G.:* Electrochemical Production of Initiators for Polymerization Processes. Vol. 38, pp. 27–54.
Slichter, W. P.: The Study of High Polymers by Nuclear Magnetic Resonance. Vol. 1, pp. 35–74.
Small, P. A.: Long-Chain Branching in Polymers. Vol. 18.
Smets, G.: Block and Graft Copolymers. Vol. 2, pp. 173–220.
Sohma, J. and *Sakaguchi, M.:* ESR Studies on Polymer Radicals Produced by Mechanical Destruction and Their Reactivity. Vol. 20, pp. 109–158.
Sotobayashi, H. und *Springer, J.:* Oligomere in verdünnten Lösungen. Vol. 6, pp. 473–548.
Sperati, C. A. and *Starkweather, Jr., H. W.:* Fluorine-Containing Polymers. II. Polytetrafluoroethylene. Vol. 2, pp. 465–495.
Sprung, M. M.: Recent Progress in Silicone Chemistry. I. Hydrolysis of Reactive Silane Intermediates. Vol. 2, pp. 442–464.
Stahl, E. and *Brüderle, V.:* Polymer Analysis by Thermofractography. Vol. 30, pp. 1–88.
Stannett, V. T., Koros, W. J., Paul, D. R., Lonsdale, H. K., and *Baker, R. W.:* Recent Advances in Membrane Science and Technology. Vol. 32, pp. 69–121.
Stille, J. K.: Diels-Alder Polymerization. Vol. 3, pp. 48–58.
Stolka, M. and *Pai, D.:* Polymers with Photoconductive Properties. Vol. 29, pp. 1–45.
Subramanian, R. V.: Electroinitiated Polymerization on Electrodes. Vol. 33, pp. 33–58.
Sumitomo, H. and *Okada, M.:* Ring-Opening Polymerization of Bicyclic Acetals, Oxalactone, and Oxalactam. Vol. 28, pp. 47–82.
Szegö, L.: Modified Polyethylene Terephthalate Fibers. Vol. 31, pp. 89–131.
Szwarc, M.: Termination of Anionic Polymerization. Vol. 2, pp. 275–306.
Szwarc, M.: The Kinetics and Mechanism of N-carboxy-α-amino-acid Anhydride (NCA) Polymerization to Poly-amino Acids. Vol. 4, pp. 1–65.
Szwarc, M.: Thermodynamics of Polymerization with Special Emphasis on Living Polymers. Vol. 4, pp. 457–495.
Tani, H.: Stereospecific Polymerization of Aldehydes and Epoxides. Vol. 11, pp. 57–110.
Tate, B. E.: Polymerization of Itaconic Acid and Derivatives. Vol. 5, pp. 214–232.
Tazuke, S.: Photosensitized Charge Transfer Polymerization. Vol. 6, pp. 321–346.
Teramoto, A. and *Fujita, H.:* Conformation-dependent Properties of Synthetic Polypeptides in the Helix-Coil Transition Region. Vol. 18, pp. 65–149.
Thomas, W. M.: Mechanism of Acrylonitrile Polymerization. Vol. 2, pp. 401–441.
Tobolsky, A. V. and *DuPré, D. B.:* Macromolecular Relaxation in the Damped Torsional Oscillator and Statistical Segment Models. Vol. 6, pp. 103–127.
Tosi, C. and *Ciampelli, F.:* Applications of Infrared Spectroscopy to Ethylene-Propylene Copolymers. Vol. 12, pp. 87–130.
Tosi, C.: Sequence Distribution in Copolymers: Numerical Tables. Vol. 5, pp. 451–462.
Tsuchida, E. and *Nishide, H.:* Polymer-Metal Complexes and Their Catalytic Activity. Vol. 24, pp. 1–87.
Tsuji, K.: ESR Study of Photodegradation of Polymers. Vol. 12, pp. 131–190.
Tuzar, Z., Kratochvíl, P., and *Bohdanecký, M.:* Dilute Solution Properties of Aliphatic Polyamides. Vol. 30, pp. 117–159.
Valvassori, A. and *Sartori, G.:* Present Status of the Multicomponent Copolymerization Theory. Vol. 5, pp. 28–58.
Voorn, M. J.: Phase Separation in Polymer Solutions. Vol. 1, pp. 192–233.
Werber, F. X.: Polymerization of Olefins on Supported Catalysts. Vol. 1, pp. 180–191.
Wichterle, O., Šebenda, J., and *Králíček, J.:* The Anionic Polymerization of Caprolactam. Vol. 2, pp. 578–595.
Wilkes, G. L.: The Measurement of Molecular Orientation in Polymeric Solids. Vol. 8, pp. 91–136.

Williams, G.: Molecular Aspects of Multiple Dielectric Relaxation Processes in Solid Polymers. Vol. 33, pp. 59–92.
Williams, J. G.: Applications of Linear Fracture Mechanics. Vol. 27, pp. 67–120.
Wöhrle, D.: Polymere aus Nitrilen. Vol. 10, pp. 35–107.
Wolf, B. A.: Zur Thermodynamik der enthalpisch und der entropisch bedingten Entmischung von Polymerlösungen. Vol. 10, pp. 109–171.
Woodward, A. E. and *Sauer, J. A.:* The Dynamic Mechanical Properties of High Polymers at Low Temperatures. Vol. 1, pp. 114–158.
Wunderlich, B. and *Baur, H.:* Heat Capacities of Linear High Polymers. Vol. 7, pp. 151–368.
Wunderlich, B.: Crystallization During Polymerization. Vol. 5, pp. 568–619.
Wrasidlo, W.: Thermal Analysis of Polymers. Vol. 13, pp. 1–99.
Yamashita, Y.: Random and Black Copolymers by Ring-Opening Polymerization. Vol. 28, pp. 1–46.
Yamazaki, N.: Electrolytically Initiated Polymerization. Vol. 6, pp. 377–400.
Yamazaki, N. and *Higashi, F.:* New Condensation Polymerizations by Means of Phosphorus Compounds. Vol. 38, pp. 1–25.
Yoshida, H. and *Hayashi, K.:* Initiation Process of Radiation-induced Ionic Polymerization as Studied by Electron Spin Resonance. Vol. 6, pp. 401–420.
Yuki, H. and *Hatada, K.:* Stereospecific Polymerization of Alpha-Substituted Acrylic Acid Esters. Vol. 31, pp. 1–45.
Zachmann, H. G.: Das Kristallisations- und Schmelzverhalten hochpolymerer Stoffe. Vol. 3, pp. 581–687.
Zambelli, A. and *Tosi, C.:* Stereochemistry of Propylene Polymerization. Vol. 15, pp. 31–60.

A. Hebeish, J. T. Guthrie
The Chemistry and Technology of Cellulosic Copolymers

1980. 91 figures, approx. 91 tables. Approx. 500 pages
(Polymers/Properties and Applications, Volume 4)
ISBN 3-540-10164-0

The driving force behind the great scientific interest in copolymer science and technology is the search for products with useful, new or interesting properties. This monograph provides an informative account of new, improved cellulosic materials and the chemistry and technology involved in their production, as well as the first detailed description of grafted and modified celluloses.
The information contained in this book will be of great value to researchers, manufacturers, but also instructors, interested in the modification of cellulosics for textiles, paper, printing, printing inks, paints, and packaging, as well as in polymerization processes and cellulose derivativization.

Electric Phenomena in Polymer Science

1979. 1 portrait. 55 figures, 20 tables. V, 174 pages
(Advances in Polymer Science, Vol. 33)
ISBN 3-540-09456-3

Contents/Information:

G. Mengoli, *Feasibility of Polymer Film Coatings Through Electroinitiated Polymerization in Aqueous Medium*
The paper deals with the feasibility of polymeric coatings onto conductive substrates by electropolymerization as a new technique of metal protection. The limits inherent in such a technique are discussed and the most promising routes for further investigation are outlined. (34 references)

R. V. Subramanian, *Electroinitiated Polymerization on Electrodes*
The article describes recent advances in electroinitiated polymerization systems concerning the formation of polymer films on metal as well as on graphite fiber electrode surfaces. The discussion considers the structure, morphology, bonding, and adhesion of the polymer on the electrodes and the relationship of these factors to potential advances in surface coatings and graphite composites. (84 references)

G. Williams, *Molecular Aspects of Multiple Dielectric Relaxation Processes in Solid Polymers*
The article outlines our current understanding of multiple relaxations observed in crystalline and amorphous solid polymers as studied using dielectric techniques. (138 references)

H. Block, *The Nature and Application of Electrical Phenomena in Polymers*
Electrical and dielectric behavior of polymers reflect macromolecular structure and motion, both in solution and the solid state. Some polymers which have special electrical properties may have commercial potential. Mention need only be made of polymer electrets, pyroelectric polymers, photo-conductive polymers as used in electroimaging, and conductive polymers to indicate the expansion of use over that of insulators. The separation of electrical behavior into "dielectric" and "bulk conductive" properties is convenient and has been followed in this review. (296 references)

Springer-Verlag
Berlin
Heidelberg
New York

Polymer Bulletin

Editors:
Prof. H.-J. Cantow, Makromolekulare Chemie, Universität Freiburg, Stefan-Meier-Strasse 31, D-7800 Freiburg, West-Germany
Prof. J. P. Kennedy, Dept. of Polymer Science, The University of Akron, Akron, OH 44325, USA
Prof. T. Saegusa, Dept. Synthetic Chemistry, Kyoto University, Kyoto, 606, Japan

Editorial Board: H. Batzer, Basel; N. Calderon, Akron, OH; S. Cesca, San Donato Milanese; P. J. Flory, Stanford, CA; J. Furukawa, Tokyo; J. E. McGrath, Blacksburg, VA; H. K. Hall, Jr., Tucson, AZ; H. H. Kausch, Lausanne; T. Kelen, Budapest; M. Kryszewski, Lódź; A. Ledwith, Liverpool; E. Maréchal, Paris-Cedex; J. Meißner, Zürich; A. Nakajima, Kyoto; G. and S. Henrici Olivé, Research Triangle Park, NC; N. A. Plate, Moscow; B. Rånby, Stockholm; C. I. Simionescu, Bucureşti; S. Sivaram, Gujarat; D. H. Solomon, Melbourne; R. Steiner, Frankfurt/M.; H. Tadokoro, Osaka; M. Takayanagi, Fukuoka; I. Uematsu, Tokyo; C. Wippler, Strasbourg; H. Zahn, Aachen

Editorial Assistant: A. Heinrich, Springer-Verlag Heidelberg

To cope with the rapid progress of polymer science, a new journal is now published characterized by emphasis on rapid publication of papers containing a most concise description of results.
The character of the new journal is between the purely archival journals of full papers and the so-called "letter journals" consisting exclusively of short communications.

Special features:
- rapid publication of papers
- no page charge
- 50 off-prints of each paper supplied free of charge

Subscription information and sample copy upon request

Send your order to your bookseller or directly to:
Springer-Verlag, Journal Promotion Dept.,
P. O. Box 105280, D-6900 Heidelberg, FRG

North America: Springer-Verlag New York Inc., Journal Sales Dept.,
44 Hartz Way, Secaucus, NJ 07094, USA

Springer International

CHEMISTRY LIBRARY